H. Goering · A. Felgenhauer · G. Lube · H.-G. Roos · L. Tobiska

Singularly Perturbed Differential Equations

Mathematical Research · Mathematische Forschung

Wissenschaftliche Beiträge
herausgegeben von der
Akademie der Wissenschaften der DDR
Institut für Mathematik

Band 13

Singularly Perturbed Differential Equations

by
H. Goering
A. Felgenhauer
G. Lube
H.-G. Roos
L. Tobiska

Singularly Perturbed Differential Equations

by
Herbert Goering
Andreas Felgenhauer
Gert Lube
Hans-Görg Roos
Lutz Tobiska

Akademie-Verlag · Berlin 1983

Verfasser:

Prof. Dr. Herbert Goering,
Dr. Andreas Felgenhauer, Dipl.-Math. Gert Lube
Dr. Hans-Görg Roos, Dr. Lutz Tobiska

Technische Hochschule "Otto von Guericke"
3010 Magdeburg
Boleslav-Bierut-Platz 5

Die Titel dieser Schriftenreihe werden vom Originalmanuskript der Autoren reproduziert.

Erschienen im Akademie-Verlag, DDR-1086 Berlin, Leipziger Str. 3-4
Lektor: Dr. Reinhard Höppner

Lizenznummer: 202 · 100/413/83
Gesamtherstellung: VEB Kongreß- und Werbedruck, 9273 Oberlungwitz
Umschlag: Rolf Kunze
Bestellnummer: 763 163 5 (2182/13) · LSV 1065
Printed in GDR
DDR 22.-- M

INTRODUCTION

In the last decade the study of asymptotic methods for approximative solution of differential equations depending on parameters has rapidly evolved. This found its expression in a series of monographs to this subject, DE GROEN (1976), ECKHAUS (1979), LIONS (1973), O'MALLEY (1974), WASILJEWA/BUTUSOW (1973, 1978), WASOW (1968). Whereas with the monograph of LIONS a certain closing of the theory of the asymptotic approximation of solutions of differential equation problems in the sense of the integral-norm was given, the achieved progress in constructing and proving uniformly valid asymptotic approximations (in the sense of the maximum norm) of the solutions of singularly perturbed elliptic and parabolic problems has not been comprehensively presented until now. The present paper aims at doing this. Here we restrict our considerations to such problems (important for applications) for which the formal global problem (also called reduced problem) is of first order.

The first chapter has proper importance. Here the basic ideas and relations are stated and formal construction methods for the asymptotic study of perturbed problems are given. The necessity to prove the asymptotic of formal solutions is emphasized. Existence and estimate theorems for elliptic and parabolic boundary value problems are given in Chapter two and their application to singularly perturbed problems is investigated. Statements concerning existence, uniqueness and regularity of the solution of the formal global problem are contained in the third chapter.

The main part of the present material is summarized in Chapter four. In order not to interrupt the representation by numerous details, we have stated the analytical properties of the treated boundary layer equations in an appendix. In 4.1, for problems of second order, the question is studied, in which subdomains the solution of the formal global problem is already an uniform asymptotic solution. For linear problems in simple cases, the asymptotic of the formal solution is constructed and proved in 4.2. and 4.3. Such simple cases appear if the solution of the formal global problem is sufficiently regular and the characteristics behave uniformly in a neighbourhood of the boundary. Two kinds of problems of second order are investigated, for which the global formal solution is not sufficiently regular. On the one hand, results are given in 4.4 in which the characteristics are tangential to the boundary of the domain and, on the other hand, special problems with turning points and singular lines are studied in 4.5. In 4.6, finally, some results of 4.2 are extended to nonlinear elliptic problems of second order.

In order not to exceed the extent of the book, we did not deal with the application of the present theory to physical and technical problems here. The reader interested in this subject can consult for instance

GOERING (1977), NAYFEH (1973), VAN DYKE (1975), COLE (1968), KEVORKIAN/COLE (1981). Likewise we have neglected aspects of the numerical treatment of singularly perturbed problems. A first monograph (DOOLAN/MILLER/SCHILDERS (1980)) to difference methods for one-dimensional problems of the kind considered here has already been published, for other problems we refer to conference materials (HEMKER/MILLER (1979) and MILLER (1980)).

For understanding the results in Chapter 2 - 4 it is not necessary to study Chapter 1 in detail. On pages 32 - 34 - beginning with formula (1.41a/b) - the reader interested in applications of the results can find a summary of the principles for the asymptotic study of partial differential equations which is independent of the other statements of Chapter 1. Then, it is possible to continue with Chapter 2.

For orientation in the book we have numbered the Lemmata, Theorems and formulas within a chapter and, additionally, equipped with the number of the corresponding chapter. Essential assumptions have been designated for instance by (H8) and refer to the corresponding chapter. In the running text we tried to dispense with references to the literature. Quotations to the literature and further remarks can be found in the comments following each chapter.

Magdeburg, Februar 1982

A. Felgenhauer H. Goering G. Lube H.-G. Roos L. Tobiska

C O N T E N T S

NOTATIONS

ε	small positive parameter
$K, K_1, \dots$	sufficiently large, ε-independent constant
$x = (x_1, \dots, x_n)$	point in R^n
$\lvert x\rvert = \sqrt{\sum_{\mu=1}^{n} x_\mu^2}$	norm in R^n
$\beta = (\beta_1, \dots, \beta_n)$	multi-index
$\lvert\beta\rvert = \sum_{\mu=1}^{n} \lvert\beta_\mu\rvert$	length of multi-index β
$\leq$	usual ordering in R or pointwise partial ordering for functions
Ω	bounded domain in R^n
$\Gamma = \partial\Omega$	boundary of Ω , in general sufficiently regular, i.e. $\partial\Omega$ can be locally described by means of a regular function and the domain Ω lies on one side of the boundary $\partial\Omega$
ν	outward directed unit normal
$D^\beta u = \dfrac{\partial^{\lvert\beta\rvert} u}{\partial x_1^{\beta_1} \cdots \partial x_n^{\beta_n}}$	derivative of the function $u: \Omega \to R$
$D_\varrho^l D_\varphi^\gamma u = \dfrac{\partial^{l+\lvert\gamma\rvert} u}{\partial\varrho^l \partial\varphi_1^{\gamma_1} \cdots \partial\varphi_{n-1}^{\gamma_{n-1}}}$	derivative of the function $u = u(\varrho, \varphi) = u(\varrho, \varphi_1, \dots, \varphi_{n-1})$
$\operatorname{dist}(x,\Gamma) = \inf\limits_{y\in\Gamma} \lvert x-y\rvert$	distance of the point x to Γ
$sgn\ \gamma$	sign of the real number γ
$C^{2m+\alpha}(\bar{\Omega}), C^{2m+\alpha, \frac{2m+\alpha}{2m}}(\bar{Q})$	spaces of regular functions (cf. 2.1)
$L^2(\Omega)$	Lebesgue space
$(\mathcal{U} \to V)$	set of all mappings of $\mathcal{U}$ into V

1. FOUNDATIONS

1.1. Asymptotic approximations and asymptotic expansions

We define asymptotic approximations for elements in a linear space U_0 which depend on a parameter $\lambda \in \Lambda$. The linear space U_0 is called basic space for the definition of an asymptotic structure. So we consider a linear space U with

$$\mathcal{U} \subset (\Lambda \to \mathcal{U}_0)$$

of elements $u = (u_\lambda)_{\lambda \in \Lambda}$ with $u_\lambda \in U_0$. In all cases which will be investigated in this book we denote by

$$\Lambda = (0, \varepsilon_0], \quad 0 < \varepsilon_0 \ll 1$$

a set of small parameters. However, the definitions and statements are also true in the case of more general sets of parameter.

Let us assume that in U_0 a topology is generated by a family of seminorms $\{p_\alpha \mid \alpha \in \hat{\alpha}\}$, that means

$$\forall \alpha \in \hat{\alpha},\ u, v \in \mathcal{U}_0,\ \mu \in \mathbb{R} \qquad \cdot\ p_\alpha(u) \geq 0$$

$$\cdot\ p_\alpha(\mu u) = |\mu|\, p_\alpha(u)$$

$$\cdot\ p_\alpha(u+v) \leq p_\alpha(u) + p_\alpha(v).$$

If the topology is separated, U_0 is usually called locally convex. But in our case it is convenient to renounce separated space (see Example 1.2(c)).

Definition 1.1. A set $\Delta \subset (\Lambda \to \mathbb{R}^+)$ of nonnegative functionals on Λ with

(i) $\forall \alpha \in \hat{\alpha},\ u \in \mathcal{U} \qquad (p_\alpha(u_\lambda))_{\lambda \in \Lambda} \in \Delta$

(ii) $\qquad \Delta + \Delta \subset \Delta$

(iii) $\mu \in \mathbb{R},\ \mu > 0 \qquad \mu \Delta \subset \Delta$

is called set of order functions if a relation $\prec$ exists in Δ with the following properties

(iv) transitivity, $\quad \delta_1 \prec \delta_2 \prec \delta_3 \Rightarrow \delta_1 \prec \delta_3 \qquad$ (1.1a)

(v) compatibility with $\leq$, $\quad \delta_1 \prec \delta_2 \leq \delta_3 \Rightarrow \delta_1 \prec \delta_3 \qquad$ (1.1b)

$\qquad \delta_1 \leq \delta_2 \prec \delta_3 \Rightarrow \delta_1 \prec \delta_3 \qquad$ (1.1c)

(vi) non - Archimedean $\quad \delta_1 \prec \delta_3,\ \delta_2 \prec \delta_3 \Rightarrow \delta_1 + \delta_2 \prec \delta_3$ ∎ $\qquad$ (1.1d)

Example 1.1. In the case of a small parameter the relations

(a) $\delta_1 \prec \delta_2 \iff \exists\, c,\ \varepsilon_1 \leq \varepsilon_0\ \forall \varepsilon \in (0, \varepsilon_1]: \quad \delta_1(\varepsilon) \leq c\, \delta_2(\varepsilon)$

(b) $\delta_1 \prec \delta_2 \iff \lim\limits_{\varepsilon \to 0} \dfrac{\delta_1(\varepsilon)}{\delta_2(\varepsilon)} = 0$

(c) $\delta_1 \prec \delta_2 \iff \exists c \ \forall \varepsilon \in (0,\varepsilon_0] : \delta_1(\varepsilon) \leq c\,\delta_2(\varepsilon)$

are in general use. ■

Definition 1.2. For each element $\delta \in \Delta$ of a set of order functions we define a Landau - set

$$O(\delta) = \{ u \in U \mid \forall \alpha \in \hat{\alpha} : p_\alpha(u) \prec \delta \} \ . \ ■$$

In the case (b) of example 1.1 the notation $o(\delta)$ instead of $O(\delta)$ is often used. With the beginning of chapter 2 we always make use of the relation (a).

According to the definitions it is possible, that the seminorms p_α also depend on $\lambda \in \Lambda$. We shall use this fact later (cf. Theorem 1.13 or section 2.1). The most important properties of Landau - sets are characterized by

Theorem 1.1.

(i) $O(\delta)$ is a linear set, that means

$$O(\delta) + O(\delta) \subset O(\delta) \ , \ \mu \in \mathbb{R} : \ \mu\, O(\delta) \subset O(\delta)$$

(ii) $\delta_1 \prec \delta_2 \quad \Longrightarrow \quad O(\delta_1) \subset O(\delta_2)$

$\delta_1 \leq \delta_2 \quad \Longrightarrow \quad O(\delta_1) \subset O(\delta_2)$.

(iii) Let $\{q_\beta \mid \beta \in \hat{\beta}\}$ be a family of seminorms generating a finer topology in U_0 than $\{p_\alpha \mid \alpha \in \hat{\alpha}\}$, that means for any $\beta \in \hat{\beta}$ there are constants $a_\alpha \in \mathbb{R}$, only a finite number of which are nonzero, such that $q_\beta \leq \sum_{\alpha \in \hat{\alpha}} a_\alpha p_\alpha$. If for both topologies Landau - sets $O_{\hat{\beta}}(\delta)$ and $O_{\hat{\alpha}}(\delta)$ are defined, the relation $O_{\hat{\beta}}(\delta) \subset O_{\hat{\alpha}}(\delta)$ is valid.

(iv) Let $\prec_1$ and $\prec_2$ be different relations in the same set of order functions Δ . According to $\prec_1$ and $\prec_2$ different Landau - sets $O_1(\delta)$ and $O_2(\delta)$, respectively, can be defined for each $\delta \in \Delta$. If $\delta_1 \prec_1 \delta_2 \Rightarrow \delta_1 \prec_2 \delta_2$ for any $\delta_{1,2} \in \Delta$, the relation $O_1(\delta) \subset O_2(\delta)$ is satisfied. ■

Proof: Let $\delta_{1,2} \in \Delta$ and $\mu_{1,2} \in \mathbb{R}$. From $\delta_1 \prec \delta$ and (1.1c) we first obtain $\mu_1 \delta_1 \prec \delta$ for $0 \leq \mu_1 < 1$ and then, because of (1.1d) the same relation for arbitrary $\mu_1 \geq 0$. Therefore (and because of (1.1d)) (i) is valid. (ii) is a direct consequence of (1.1a,c). From (1.1c) and $p_\alpha(u) \prec \delta \Rightarrow \sum a_\alpha p_\alpha(u) \prec \delta$ it follows (iii). The proof of (iv) is evident. ■

Definition 1.3. $v \in U$ is called an asymptotic approximation of $u \in U$ if $u \in v + O(\delta)$. ■

Example 1.2. Let Ω be a n-dimensional bounded domain. In the case of $U_o = C(\bar{\Omega})$ different possibilities for the choice of seminorms produce different conceptions for asymptotic approximations. The most important of these conceptions are

(a) uniform (asymptotic) approximation

$$\hat{\alpha} = \{1\}, \quad p_1(u) = \|u\| = \sup_{x \in \Omega} |u(x)| \quad .$$

(b) weighted (asymptotic) approximation

Let $\forall \varepsilon \in \Lambda : \varphi_\varepsilon \in C(\bar{\Omega})$. Then

$$\hat{\alpha} = \{1\}, \quad p_1(u) = \|\varphi_\varepsilon u\| \quad .$$

(c) (asymptotic) approximation on a subset $\Omega^* \subset \Omega$

$$\hat{\alpha} = \{1\}, \quad p_1(u) = \|u\|^{\Omega^*} = \sup_{x \in \Omega^*} |u(x)| \quad .$$

To continue the concept of asymptotic approximation let us assume that $\bar{\Gamma} \subset \bar{\Omega}$ is a subset consisting of "boundary layers" and "limit sets", respectively, that means

$$\Gamma = \bigcup_{i=1}^{k} \Gamma_i \quad , \quad \dim \Gamma_i = q_i < n$$

Γ_i homeomorphic to an open q_i - dimensional cuboid (for $q_i > 0$).

In dependence on Γ it is possible to define

(d) global (asymptotic) approximation

$$\hat{\alpha} = (0, \alpha_o], \quad p_\alpha(u) = \|u\|^{\Omega_\alpha} = \sup_{x \in \Omega_\alpha} |u(x)|$$

with

$$\Omega_\alpha = \{ x \in \Omega \mid dist(x, \Gamma) \geq \alpha \} \quad .$$

(e) local (asymptotic) approximation

For an arbitrary $\delta \in \Delta$ with $\lim_{\varepsilon \to 0} \delta(\varepsilon) = 0$ we choose

$$\hat{\alpha} = \{1\}, \quad p_1(u) = \|u\|^{\Omega_\varepsilon^*} = \sup_{x \in \Omega_\varepsilon^*} |u(x)|$$

with $\Omega_\varepsilon^* = \{ x \in \Omega \cup \Gamma \mid dist(x, \Gamma) < \delta(\varepsilon) \}$. ■

A similar terminology can be introduced if other norms are considered. So it is possible to make use of derivatives of function (see, for instance, the ε-dependent norm $|u|_{2m+\alpha}$ in $C^{2m+\alpha}(\bar{\Omega})$ in section 2.1). In the same way integral-norms or seminorms of a locally convex space can be taken as a basis.

In the following we shall define asymptotic expansions.

Definition 1.4. A sequence $(\delta_\mu)_{\mu=1}^{N}$ and $(\delta_\mu)_{\mu=1}^{\infty}$, respectively with $\delta_\mu \in \Delta$, $\delta_\mu \prec \delta_{\mu+1} \nprec \delta_\mu$ is called a scale of magnitudes. Additionally, we require

$$\delta_\mu(\varepsilon) > 0 , \quad \lim_{\varepsilon \to 0} \frac{\delta_{\mu+1}(\varepsilon)}{\delta_\mu(\varepsilon)} = 0 \quad . \quad ■$$

This implies

$$O(\delta_{\mu+1}) \subset O(\delta_\mu) \tag{1.2}$$

Definition 1.5. For any scale of magnitudes an asymptotic structure is defined by the set

$$\omega = \{ O(\delta_\mu) \mid \mu = 1,2,\ldots \} \quad . \quad ■$$

In this way also non-denumerable asymptotic structures can be defined (for instance with the help of gauge sets of order functions, see ECKHAUS (1977b)). We do not consider this in detail. Furthermore, in the first chapter we shall frequently take N = 1. This is no restriction of generality for theoretical investigations in the case $N < \infty$ because the cases $\omega = \{ O(\delta_\mu) \mid \mu = 1, \ldots, N\}$ and $\omega = \{ O(\delta_\mu) \mid \mu = N\}$ yield equivalent results.

Definition 1.6. In a space U with an asymptotic structure $\omega = \{ O(\delta_\mu) \mid \mu = 1,2,\ldots\}$ we define

$\sum_{\mu \geq 0} u_\mu$ to be a formal series, if $u_0 \in U$, $u_\mu \in O(\delta_\mu)$, $(\mu \geq 1)$,

$\sum_{\mu \geq 0} u_\mu$ to be an asymptotic expansion of $u \in U$ if

$$u \in \sum_{\mu=0}^{K} u_\mu + O(\delta_{K+1}) \tag{1.3}$$

- then u is called the asymptotic sum of series - and u to be asymptotically zero if $u \in O_\omega = \bigcap_{\mu \geq 1} O(\delta_\mu)$. ■

Definition 1.7. A space U with an asymptotic structure ω is called asymptotically complete if any formal series has (at least) one asymptotic sum. ■

According to the definition it follows that any asymptotic expansion is a formal series. Furthermore an asymptotic sum is exactly determined except for a difference being asymptotically zero.

By the pair (U,ω) we defined a space with asymptotic structure in which asymptotic approximations und expansions can be realized in natural way. It is briefly denoted by

$$as_\Lambda(U_0, \{p_\alpha\}; \Delta) \quad .$$

The essential point is the range of the set U considered as a subset of $(\Lambda \to U_0)$ (see section 1.2). If it is obvious which seminorms are used (for instance the norm in the space U_0) and the set of parameters is fixed we shall write

$$as_\Lambda(U_0; \Delta), \quad as(U_0, \{p_\alpha\}; \Delta) \quad and \quad as(U_0, \Delta),$$

respectively. For simplicity we permit to write instead of the sets $\{p_\alpha\}$ and Δ, respectively their elements.

1.2. Expansion operators

Let $U = as(U_0, \Delta)$ be a given space with asymptotic structure. For the construction of an asymptotic expansion of an explicitly or implicitly given element $u \in U$ the definition 1.6. is not convenient because of u_μ, $\mu = 0,1,\dots,k$ from (1.3) are not uniquely determined (This problem was already mentioned by ERDELYI (1956)). For that reason we now introduce the concept of expansion operators. In general we restrict our considerations to the case of an asymptotic structure containing exactly one Landau - set $\omega = \{O(\delta)\}$, therefore to the case $U = as(U_0; \delta)$.

Definition 1.8. A linear operator E mapping a linear space into itself is called projector if EE = E. The kernel of a projector is denoted by

$$O(E) = \{ u \in U \mid Eu = 0 \} \quad . \quad ■$$

Conclusion. (Approximation by projectors)
For projectors E it is valid that

$$u \in Eu + O(E) \qquad \text{for every } u \in U \quad . \quad ■$$

Definition 1.9. A projector E is called regular expansion operator in $as(U_0; \delta)$ if

$$O(E) = O(\delta) \quad . \quad ■$$

In the case of more general sets $\omega = \{O(\delta_\mu) \mid \mu = 0,1,\dots\}$ we consider adequate sequences of projectors $(E^\mu)_{\mu \geq 0}$ satisfying the relations

$$O(E^\mu) = O(\delta_{\mu+1}) \quad .$$

It follows (because of (1.2))

$$E^\mu E^{\mu+1} = E^\mu \quad . \qquad (1.4)$$

Additionally, in this case we assume

$$E^{\mu+1} E^\mu = E^\mu \quad .$$

The validity of this relation can always be realized by transition to the expansion operators $\bar{E}^{\mu+1} = E^{\mu+1} + E^\mu - E^{\mu+1}E^\mu$ with

$O(E^{\mu+1}) = O(E^{\mu+1})$ (see Theorem 1.3).

Definition 1.10. A projector E is called expansion operator if

$O(E) \supset O(\delta)$, that means $E\, O(\delta) = \{0\}$. ■

Definition 1.11. An expansion operator E is called regular at the point $u \in U$ if

$u \in E\, u + O(\delta)$.

It is called regular on the subset $V \subset U$ if it is regular at u for every $u \in V$. ■

Conclusion. An expansion operator E is regular on the linear subspace $EU + O(\delta)$. ■

Hence, regular expansion operators define an asymptotic approximation in $as(U_0;\Delta)$ for every $u \in U$, whereas, in general, expansion operators define formal series the asymptotic sum of which is not necessarily u.

If it is not immediately possible to construct a regular expansion operator, we can try to construct a new one from an appropriate family (E_κ) of expansion operators.

Definition 1.12. A projector E is called composite expansion operator of a given family of expansion operators (E_κ) if $O(E) = \bigcap_\kappa O(E_\kappa)$ and

$$E_\kappa E = E \tag{1.5}$$

$$O(E) \supset \bigcap_\kappa O(E_\kappa), \tag{1.6}$$

respectively. ■

Conclusion. E is regular on the set $V = E(U) + O(\delta)$ and $E_\kappa V = E_\kappa U$ is valid. ■

We now introduce the concrete expansion operators being important for the asymptotic investigation of singularly perturbed partial differential equations (see also Example 1.2.). Let $\Omega \subset \mathbb{R}^n$ be a n-dimensional domain and let $\Gamma_i \subset \bar{\Omega}$ be connected q_i-dimensional subsets ($q_i < n$), which are homeomorphic to open q_i -dimensional cuboids for $q_i > 0$. Here $\Gamma = \bigcup_{i=1}^{\kappa} \Gamma_i$ represents the union of occurring "boundary layers" and "limit sets", respectively.

Definition 1.13. Let $(\delta_\mu)_{\mu\geq 0}$ be a scale of magnitudes. Global expansion operators $E^\mu(\delta_0,\delta_1,\dots,\delta_\mu)$ (or simple E^μ if no confusion is to expect) are defined by

$$E^0(\delta_0) = E(\delta_0)$$

$$E^\mu(\delta_0,\dots,\delta_\mu) = E^{\mu-1}(\delta_0,\dots,\delta_{\mu-1}) + E(\delta_\mu)(I - E^{\mu-1}(\delta_0,\dots,\delta_{\mu-1}))$$

where

$$(E(\delta)u)(x) = \delta \lim_{\varepsilon \to 0} \delta^{-1} u(x), \quad x \in \Omega \setminus \Gamma .$$

The values of the function $E^{\mu}u$ at a point $x \in \Gamma$ are obtained from the regularity of the basic space U_o. ∎

Similarly, local expansion operators can be defined for the several Γ_i. For simplicity we omit the index i. We assume that for $\Gamma = \Gamma_i$ sufficient regular local coordinates $y = (\rho, \varphi)$ with $\rho = (\rho_1, \ldots, \rho_{n-q})$ and $\varphi = (\varphi_1, \ldots, \varphi_q)$ exist such that $x \in \Gamma$ if and only if $\rho = 0$. In dependence on real, monotonic increasing positive functions η_μ of $\varepsilon \in \Lambda$ $(\mu = 1, \ldots, n-q)$ we construct an operator with

$$(E(\Gamma; \eta_1, \ldots, \eta_{n-q}; \delta) u)(\rho, \varphi) = \delta \lim_{\substack{\varepsilon \to 0 \\ \rho_\mu / \eta_\mu = const}} \delta^{-1} u(\rho, \varphi) \quad \text{for } \rho > 0$$

The limit has to be interpreted in such a way that first the ρ-coordinates are "stretched" by $\rho_\mu = \eta_\mu(\varepsilon) \zeta_\mu$ and then ε tends to zero under the condition that $\zeta = (\zeta_1, \ldots, \zeta_{n-q})$ is independent of ε.

__Definition 1.14.__ Let $(\delta_\mu)_{\mu \geq 0}$ be a scale of magnitudes and let $(\eta_\mu)_{\mu=1}^{n-q}$ be real monotonic increasing functions on Λ. __Local expansion operators__ $E^{\mu}(\Gamma; \eta_1, \ldots \eta_{n-q}; \delta_0, \delta_1, \ldots, \delta_\mu)$ (or $E^{\mu}_{\eta_1, \ldots, \eta_{n-q}}$)

are defined by

$$E^{0}(\Gamma; \eta_1, \ldots, \eta_{n-q}; \delta_0) = E(\Gamma; \eta_1, \ldots, \eta_{n-q}; \delta_0)$$

$$E^{\mu}(\Gamma; \eta_1, \ldots, \eta_{n-q}; \delta_0, \ldots, \delta_\mu) = E^{\mu-1}(\Gamma; \eta_1, \ldots, \eta_{n-q}; \delta_0, \ldots, \delta_{\mu-1}) + E(\Gamma; \eta_1, \ldots, \eta_{n-q}; \delta_\mu)(I - E^{\mu-1}(\Gamma; \eta_1, \ldots, \eta_{n-q}; \delta_0, \ldots, \delta_{\mu-1})).$$

The values of the function at a point $x \in \Gamma$ are again obtained from the regularity of the basic space U_o. ∎

In such a way the operators $E^{\mu}_{\eta_1, \ldots, \eta_{n-q}}$ are defined only in a neighbourhood of Γ. For theoretical investigations we assume that the operators can be extended to the whole domain Ω. The problems connected with that are unimportant for application.

In the concrete application the choice of order functions δ_μ and η_μ is essentially depended on the fact that the considered $u \in U$ lie in the domain of definition of the expansion operator but they do not so in its kernel.

The expansions defined by local expansion operators have the structure

$$(E^{k}_{\eta_1, \ldots, \eta_{n-q}} u)(x) = \sum_{\mu=0}^{k} a_\mu(x) \delta_\mu$$

with

$$a_\mu(x) = b_\mu\left(\frac{\varrho_1}{\eta_1}, \dots, \frac{\varrho_{n-q}}{\eta_{n-q}}, \varphi_1, \dots, \varphi_q\right).$$

For constructing expansion operators composed by global and local expansion operators which then are regular in many cases the concept of matched asymptotic expansions has proved its worth. Here we formulate it for general expansion operators.

Definition 1.15. Let

$$E = E(E_1, \dots, E_k) \tag{1.7}$$

be a mapping such that k projectors on U are mapped on a linear operator E mapping U in U. If E is a composite expansion operator for any family of expansion operators $(E_i)_{i=1}^{k}$ satisfying a formal relation

$$M(E_1, \dots, E_k) = 0$$

we call this relation a matching rule. E is the associated composite expansion operator. ■

Now, we give important examples for matching-rules and composite expansion operators.

Theorem 1.2. Let Z_i $(i=1,\dots,k)$ be linear operators mapping U in U. The operator

$$E = \sum_{i=1}^{k} Z_i E_i \tag{1.8}$$

is a composite expansion operator of $(E_i)_{i=1}^{k}$ if and only if the matching rule

$$\sum_{i=1}^{k} E_j Z_i E_i = E_j \quad , \quad (j = 1, \dots, k) \tag{1.9}$$

is satisfied. ■

Proof. (1.9) is equivalent to (1.5) for the operator (1.8). From $E_i u = 0$ $(i=1,\dots,k)$ it follows that $Eu = \sum_{i=1}^{k} Z_i 0 = 0$. It remains to show that E is a projector. This can be obtained from (1.5)

$$E E = \sum_{i=1}^{k} Z_i E_i E = \sum_{i=1}^{k} Z_i E_i = E \quad . \quad ■$$

Theorem 1.3. The operator

$$E = E_1 + E_2 - E_2 E_1 \tag{1.10}$$

is a composite expansion operator of E_1, E_2 if and only if the matching rule

$$E_1(E_2 E_1 - E_1 E_2) = 0 \tag{1.11}$$

is satisfied. The matching rules

$$E_2 E_1 - E_1 E_2 = 0 \tag{1.12}$$

and

$$E_1 - E_1 E_2 = O, \tag{1.13}$$

respectively, are sufficient for (1.11). ■

<u>Proof.</u> The theorem is a special case of the preceding one (where $Z_1 = I - E_2$ and $Z_2 = I$). By multiplying (1.12) and (1.13), respectively from the left and right, respectively we obtain (1.11). ■

<u>Theorem 1.4.</u> Let E' be a projector. Then the relations

$$E_1(E' - E_2) = 0\,, \qquad E_2(E' - E_1) = 0 \tag{1.14}$$

$$E'(E' - E_2) = 0\,, \qquad E'(E' - E_1) = 0 \tag{1.15}$$

are matching-rules with the associated composite operator

$$E = E_1 + E_2 - E' \quad . \tag{1.16}$$

■

<u>Proof.</u> Because of (1.15) E can be written in the form $E = E_1 + (I-E')E_2$ and Theorem 1.2 is applicable. ■

In general, matching-rules are not satisfied on the whole considered space. The range of validity, that means the linear subset on which E is a composite expansion operator, is a subset of $U_M = \{u \in U \mid M(E_1,\dots,E_K)u = 0\}$. It includes every set $V \subset U_M$ for which the steps of the formal proof of matching rule do not take out. For the matching-rules (1.9), (1.11), (1.12) and (1.14)/(1.15) the range of validity is identical with U_M.

Now we consider a family of expansion operators on U:

$$\mathcal{E} = \{ E_\gamma \mid \gamma \in \hat{\gamma} \} \ .$$

<u>Definition 1.16.</u> (ECKHAUS (1973)). $E_1 \in \mathcal{E}$ <u>is contained in</u> $E_2 \in \mathcal{E}$ if and only if $E_1 = E_1 E_2$. ■

<u>Conclusion.</u> E_1 is contained in E_2 if and only if E_2 is a composite expansion operator of $\{E_1, E_2\}$ that means if and only if $O(E_2) \subset O(E_1)$. ■

<u>Definition 1.17.</u> Two expansion operators E_1, $E_2 \in \mathcal{E}$ are called <u>equivalent</u> if $E_1 = E_1 E_2$, $E_2 = E_2 E_1$. ■

<u>Definition 1.18.</u> An expansion operator $E_1 \in \mathcal{E}$ <u>is significant</u> (with respect to $\mathcal{E}$) if E_1 is contained only in equivalent expansion operators. ■

<u>Conclusion.</u> The statement "E is composite expansion operator of $\mathcal{E}$" is equivalent to

(i) E contains every $E_\gamma \in \mathcal{E}$

(ii) E is contained in any E' containing every $E_\gamma \in \mathcal{E}$.

Particularly, E is equivalent to any significant expansion operator in $\mathcal{E} \cup \{E\}$. ■

Hence for the construction of composite operators we can neglect all the expansion operators which are not significant themselves but are contained in a significant one.

In many cases it is senseless (or impossible) to exactly characterize the set U in order to verify a relation of the form $E_1 = E_1 E_2$ exactly. Therefore we define, for $E_1, E_2 \in \mathcal{E}$

Definition 1.19. E_1 is contained in E_2 at the point $u \in U$ if $E_1(I-E_2)u=0$. E_1 and E_2 are equivalent at the point $u \in U$ if $E_1(I-E_2)u = E_2(I-E_1)u=0$. E_1 is called significant with respect to $\mathcal{E}$ at the point $u \in U$ if at this point E_1 is contained only in equivalent expansion operators. ■

With the help of matching-rules and composite expansion operators for two or more than two expansion operators a projector can be constructed which approximates "just as well" as the whole considered family. However the composite expansion operator is not necessarily regular. For this we need additional assumptions. For instance, we can formulate the following

Lemma 1.1. Let E_1, E_2 be two expansion operators in a space U with asymptotic structure $\{O(\delta)\}$ and let U_1 and U_2 be two subspaces such that

E_i regular on U_i $(i=1,2)$

$$E_1 U \subset U_1 \tag{1.17a}$$

$$E_2\, O(E_1) \subset U_2 \subset O(E_1)\,. \tag{1.17b}$$

Then we have $E_1E_2 = E_1E_2E_1$ on U and according to Theorem 1.3 the composite expansion operator $E = E_1 + E_2 - E_2E_1$ is regular on $U_1 + U_2$. ■

Proof. The matching-rule follows directly from (1.17b). For $u = u_1 + u_2$ with $u_1 \in U_1$ and $u_2 \in U_2$ it follows from (1.17a) that

$$E_1 u = E_1 u \in u_1 + O(\delta)$$

$$E_2 E_1 u = E_2 u_1$$

$$E_2 u = E_2 u_1 + E_2 u_2 \in E_2 u_1 + u_2 + O(\delta)\,.$$

Summing up we obtain $Eu \in u + O(\delta)$. ■

Conclusion. Let Z be a linear operator satisfying the relations

$$E_1 Z = 0\,, \qquad E_2 Z = I \tag{1.18}$$

on the set $E_2 O(E_1)$. Then

$$\bar{E} = E_1 + Z E_2 (I - E_1) \tag{1.19}$$

is another expansion operator (equivalent to E) regular on $U_1 + U_2$. ∎

<u>Proof.</u> Applying Theorem 1.2 we obtain

$$E_1\bar{E} = E_1 + E_1 Z E_2(I-E_1) = E_1 \; ,$$

$$E_2\bar{E} = E_2E_1 + E_2 Z E_2(I-E_1) = E_2E_1 + E_2(I-E_1) = E_2 \; .$$ ∎

<u>Remark.</u> The relation (1.17b) is equivalent to

$$u_2 \in \mathcal{U}_2 \quad \Longrightarrow \quad E_1 u_2 = 0 \tag{1.20}$$

$$\text{and} \quad u \in \mathcal{U} \quad \Longrightarrow \quad E_2(I-E_1)u \in \mathcal{U}_2 \; .$$

In the case of $U = U_1 + U_2$ the second condition is equivalent to $E_2U_2 \subset U_2$. Assuming, without restriction of generality, that $U_i = U_i + O(\delta)$, $i = 1,2$, we can guarantee $E_iU_i \subset U_i$, $i=1,2$, such that the essential point of the assumptions (1.17a,b) is only (1.20). ∎

In the asymptotic treatment of singularly perturbed differential equations a situation as supposed in Lemma 1.1 will result in success. For illustration we shall outline the case of a simple boundary layer of a function of one variable. Let Ω be an interval and $\Gamma = \{0\} \subset \bar{\Omega}$. The following considerations can also be transferred to corresponding boundary layer structures for $\dim \Omega \geq 1$. We consider global and local expansion operators

$$E_1^{\mu} = E^{\mu}(\delta_0, \delta_1, \ldots, \delta_{\mu})$$

$$E_2^{\mu} = E^{\mu}(\Gamma; \eta; \delta_0, \delta_1, \ldots, \delta_{\mu}) \; .$$

For simplicity we suppose that both expansions are based on the same scale of magnitudes. Otherwise it is necessary to complete the expansions by zeros.

<u>Definition 1.20.</u> In dependence on η and $(\delta_{\mu})_{\mu \geq 0}$ a function $v(x,\varepsilon)$ is called a <u>boundary layer function</u> for the boundary layer $\Gamma = \{0\}$ if real monotonic increasing functions h_{μ} exist with

$$\lim_{\varepsilon \to 0} \frac{\delta_{\mu}(\varepsilon)}{\delta_{K+1}(\varepsilon)}\, h_{\mu}(t\,\eta(\varepsilon)) = 0 \, , \qquad 0 \leq \mu \leq K+1 \, , \; t > 0 \tag{1.21}$$

such that the estimates

$$\begin{aligned} |v(x,\varepsilon)| &\leq h_0\left(\tfrac{\eta}{x}\right)\delta_0(\varepsilon) \, , & \mu = 0 \\ |v(x,\varepsilon) - (E_2^{\mu-1}v)(x,\varepsilon)| &\leq h_{\mu}\left(\tfrac{\eta}{x}\right)\delta_{\mu}(\varepsilon) \, , & 1 \leq \mu \leq K+1 \end{aligned} \tag{1.22}$$

are valid. ∎

Let now U_1 be a subspace of $as(C(\bar{\Omega}); \delta_0, \ldots, \delta_{K+1})$ in which E_1^{K+1} is defined and E_1^{K} is regular. For U_2 we choose a linear set of boundary functions. Thereby we can suppose $E_iU_i \subset U_i$, $i = 1,2$. Furthermore let $\psi \in C^{\infty}[0,\infty)$ be a cut off function with

$$\psi(t) = \begin{cases} 1 & 0 \le t \le 1 \\ 0 & t \ge 2 \end{cases} .$$

Lemma 1.2. On the set $U = U_1 + U_2$ the matching-rule

$$E_1^\kappa E_2^\kappa E_1^\kappa = E_1^\kappa E_2^\kappa$$

is valid. Composite regular expansions for u are given by

$$E^\kappa u = E_1^\kappa u + v$$

and by

$$(E^\kappa u)(x,\varepsilon) = (E_1^\kappa u)(x,\varepsilon) + \psi\left(\frac{x}{\tilde{\eta}(\varepsilon)}\right) v(x,\varepsilon)$$

with $v = E_2^k (u - E_1^\kappa u)$, provided that $\tilde{\eta}(\varepsilon)$ is a strong monotonic increasing function with

$$\lim_{\varepsilon \to 0} \tilde{\eta}(\varepsilon) = 0 , \qquad \lim_{\varepsilon \to 0} \frac{\eta(\varepsilon)}{\tilde{\eta}(\varepsilon)} = 0 \qquad . \qquad ■$$

Proof. From the construction of E_2^κ it follows the existence of a monotonic increasing function $M_{\kappa+1}(t)$ such that for boundary layer functions the estimate

$$|v(x,\varepsilon) - (E_2^\kappa v)(x,\varepsilon)| \le M_{\kappa+1}\left(\frac{x}{\eta(\varepsilon)}\right) \delta_{\kappa+1}(\varepsilon)$$

is valid. From (1.22) for $\mu = K + 1$ and because of

$$\tilde{K} = \max_{0<t<\infty} \left(\min\{M_{\kappa+1}(t), h_{\kappa+1}(\tfrac{1}{t})\}\right) < \infty$$

we obtain the relation

$$|v(x,\varepsilon) - (E_2^\kappa v)(x,\varepsilon)| \le \tilde{K}\, \delta_{\kappa+1}(\varepsilon) ,$$

that means E_2^κ is regular for boundary layer functions. Furthermore, from (1.22) and (1.21) for $\mu = 0$ according to the definition of E_1^κ it follows that $E_1^\kappa v = 0$ for boundary layer functions v. All assumptions in Lemma 1.1 are fulfilled, where the multiplication with the cut off function $\psi(\frac{x}{\tilde{\eta}})$ corresponds to a linear operator satisfying the conditions (1.18). Lemma 1.2 is a consequence of Lemma 1.1 and the conclusion. ■

Remark. It is usual to consider the limits (1.21) not for $\varepsilon \to 0$ but for $\xi = \frac{1}{t\eta(\varepsilon)} \to \infty$. In application this are the decay conditions at infinity for the remainders in (1.22). In the case of a power series it is well-known that the remainders can be estimated by the derivatives of the function $v^{\circ}(\xi, \mathcal{E}) = v(\eta(\mathcal{E})\xi, \mathcal{E})$. ■

Another approach to the regularity of composite expansions can be found by using the overlap hypothesis. Here a third family of expansion operators (intermediate expansion operators)

$$E'^{\mu} = E^{\mu}(\Gamma; \tilde{\eta}; \delta_0, \delta_1, \ldots, \delta_{\mu})$$

is considered with

$$\lim_{\varepsilon \to 0} \tilde{\eta}(\varepsilon) = 0, \qquad \lim_{\varepsilon \to 0} \frac{\eta(\varepsilon)}{\tilde{\eta}(\varepsilon)} = 0 \quad :$$

Matching-rule and composite expansion have the form given in Theorem 1.4. The formulas (1.15) represent the overlap hypothesis. See also ECKHAUS (1973), BERG (1976).

1.3. Asymptotic solutions

In what follows we shall consider problems of the form

$$L_{\varepsilon} u_{\varepsilon} = f_{\varepsilon}, \qquad x \in \Omega \subset \mathbb{R}^n, \qquad \varepsilon \in \Lambda = (0, \varepsilon_0], \tag{1.23a}$$

$$l_0 u_{\varepsilon} = 0, \qquad x \in \Gamma_{-} \subset \partial\Omega, \tag{1.23b}$$

$$l_1 u_{\varepsilon} = g_{\varepsilon}, \qquad x \in \Gamma' \subset \partial\Omega, \tag{1.23c}$$

where

$$L_{\varepsilon} = \varepsilon L_1 + L_0, \qquad L_{\varepsilon} \in (\mathcal{U}_0 \to V_0^{(1)}), \; l_1 \in (\mathcal{U}_0 \to V_0^{(2)}) .$$

L_0, L_1 denote differential operators over the domain Ω. For simplicity we assume that the boundary conditions are divided on l_0 and l_1 in such a way that (1.23) and the global problem

$$L_0 u_0 = f_0, \qquad x \in \Omega$$

$$l_0 u_0 = 0, \qquad x \in \Gamma_{-} \subset \partial\Omega$$

are correctly formulated and that l_0 is linear. This enables us to consider U_0 to be chosen in such a way that every element of U_0 satisfies the boundary condition $l_0 u = 0$.

Let

$$u = (u_{\varepsilon})_{\varepsilon \in \Lambda}, \qquad f = (f_{\varepsilon}, g_{\varepsilon})_{\varepsilon \in \Lambda}$$
$$A u = (L_{\varepsilon} u_{\varepsilon}, l_1 u_{\varepsilon})_{\varepsilon \in \Lambda} \tag{1.24}$$

and let $U = as(U_0, \Delta_u)$ and $V = as(V_0, \Delta_v)$ be appropriate spaces with asymptotic structures ω_u and ω_v on the basic spaces U_0 and $V_0 = V_0^{(1)} \times V_0^{(2)}$, respectively. Then the problem (1.23) can be considered as a special case of

$$A u = f \,, \qquad u \in \mathcal{U} \,, \qquad A \in (\mathcal{U} \to V) \,. \tag{1.25}$$

In detail, we only study the case of asymptotic structures with exactly one Landau-set $\omega_{\mathcal{U}} = \{0_{\mathcal{U}}\}$ and $\omega_V = \{0_V\}$, respectively. Definitions and statements can be transferred to asymptotic structures with a finite or denumerable number of elements in an analogous way. See also the remarks following definition 1.5.

Now we consider the problem (1.25) in more detail .

Definition 1.21. For the problem (1.25) we call

$u^* \in \mathcal{U}^* = A^{-1}\{f\}$ exact solution,

$\tilde{u} \in \tilde{\mathcal{U}} = A^{-1}\{f\} + O_{\mathcal{U}}$ asymptotic solution and

$\bar{u} \in \bar{\mathcal{U}} = A^{-1}\{f + O_V\}$ strong formal solution. ■

Definition 1.22. The operator $A \in (U \to V)$ is called asymptotically continuous at the point $u \in U$ if

$$A(u + O_{\mathcal{U}}) \subset Au + O_V \quad .$$

■

Theorem 1.5. If A is continuous for every strong formal solution $\bar{u} \in \bar{U}$, then we have $v \in \bar{U}$ for every $v \in \bar{u} + O_{\mathcal{U}}$, especially

$$\tilde{\mathcal{U}} \subset \bar{\mathcal{U}}$$

and every asymptotic solution is a strong formal solution. ■

Proof. It is true, that

$$Av \in A(\bar{u} + O_{\mathcal{U}}) \subset A\bar{u} + O_V = f + O_V$$

for $v \in \bar{u} + O_{\mathcal{U}}$. Hence, we obtain $v \in \bar{U}$. ■

Now, let E and E', respectively be given expansion operators in U and V.

Definition 1.23. A pair (E,E') of expansion operators in U and V, respectively is called asymptotic method for the problem (1.25). ■

Applying an asymptotic method we consider the problem

$$E'Au = E'f \,, \qquad u \in E\mathcal{U} \,, \tag{1.26}$$

instead of (1.25).

Definition 1.24. A solution of (1.26)

$$\hat{u} \in \hat{\mathcal{U}} = (E'A)^{-1}\{E'f\} \cap E\mathcal{U}$$

is called weak formal solution of (1.25). ■

__Definition 1.25.__ An asymptotic method is called __consistent at a point__ $u \in U$ (for an operator A) if from $E(u-v) = 0$ it follows that $E'(Au-Av) = 0$ for every $v \in U$. It is called __consistent__ if it is consistent for every $u \in U$. ■

__Conclusion.__ If an asymptotic method is consistent at the point $u \in U$, then the relation

$$E'AEu = E'Au$$

is valid. ■

A direct consequence from the definitions is the following

__Theorem 1.6.__ If A is asymptotically continuous and E is a regular expansion operator (at the point $u \in U$), then the asymptotic method (E,E') is consistent (at $u \in U$). ■

__Proof.__ From the asymptotic continuity it follows that in correspondence with the definition of expansion operators (see Definition 1.10)

$$E'A(u+O_u) = \{E'Au\} \quad .$$

From the regularity of E we obtain $Eu \in u + O_u$ and consequently, the asymptotic method is consistent . ■

__Theorem 1.7.__ If the asymptotic method (E,E') is consistent at $u^* \in U^*$ and $\bar{u} \in \bar{U}$, respectively, then it holds that $Eu^* \in \hat{U}$ and $E\bar{u} \in \hat{U}$, respectively. Hence, in the case of a consistent method every strong formal solution (and, consequently, also every exact solution) is a weak formal solution. ■

__Proof.__ Because of $U^* \subset \bar{U}$ it is sufficient to prove the second case. It follows that

$$E'AE\bar{u} = E'A\bar{u} \in E'(f+O_V) = \{E'f\} \quad . \quad ■$$

__Theorem 1.8.__ Let (E,E') be an asymptotic method and let E' be a regular expansion operator. Then every weak formal solution is a strong formal solution, such that $\hat{U} \subset \bar{U}$. If the asymptotic method additionally is consistent it follows that

$$\hat{U} = E\bar{U} \tag{1.27a}$$

$$\bar{U} = \hat{U} + O(E) \quad . \quad ■ \tag{1.27b}$$

Therefore, in the case of a consistent method (E,E'), where E' is regular, the terminus weak formal solution coincides with the terminus formal solution in the sense of (1.27). Then we shall simply use the terminus __formal solution__ of (1.25).

__Proof of Theorem 1.8.__ For $\hat{u} \in \hat{U}$ it follows that $E'A\hat{u} = E'f$, consequently, since E' is regular, we obtain $A\hat{u} \in f + O_V$ and $\hat{u} \in \bar{U}$, respectively. From that and Theorem 1.7 in the case of consistent methods we have

$E\bar{U} \in \hat{U} \subset \bar{U}$ and from $\hat{U} \subset EU$ it follows $\hat{U} = E\bar{U}$. Furthermore, if for $\bar{u} \in \bar{U}$ it is true that $Ev = E\bar{u}$, the relations

$$E'Av = E'AEv = E'AE\bar{u} = E'A\bar{u} = E'f$$

are satisfied that means $v \in \bar{U}$ and so we obtain $\bar{U} = \bar{U} + O(E)$, which completes the proof. ■

The application of asymptotic methods aims at constructing asymptotic solutions. However, actually, weak formal solutions have been constructed by means of (1.26). Appropriately choosing O_V and E' it is often possible to guarantee the properties of consistency and regularity such that Theorem 1.8 is applicable.

If we assume, furthermore, that the operator A is asymptotically continuous it follows from Theorem 1.5 that the set of asymptotic solutions of (1.25) is contained in the set of formal solutions. However we have to take into consideration that we consider only such asymptotic solutions which are contained in our basic set U. Consequently, we have constructed the associated expansions $E\tilde{u} \in \hat{U}$ for these asymptotic solutions $\tilde{u} \in \tilde{U}$. If the expansion operator E is regular at the point $\tilde{u}$, it can be concluded that $E\tilde{u} + O_U \subset \tilde{U}$. The regularity of E is also necessary for the conclusion that in general formal solutions are asymptotic solutions.

Up to now it is an open problem for general classes of singularly perturbed boundary value problems and given basic space $\mathcal{U}_0$ explicitly to define the space $as(U_0, \Delta)$ (that means the sets U and O_U) in connection with the expansion operators E, E' such that regularity of E, consistency and (in dependence on O_V) asymptotic continuity as well as regularity of E' are satisfied but all the interesting solutions of the problem are contained in $as(U_0, \Delta)$. Therefore it is necessary to work partially with hypothesises and to prove a-posteriori that an asymptotic solution has been constructed.

For this purpose a sufficiently complicated expansion operator E is chosen which is assumed to be regular for all solution points taken into consideration (see section 1.5). Then $U \subset (\Lambda \to U_0)$ is determined such that E is regular. The asymptotic structure and the expansion operator in the image space are defined in such a way that E' is consistent and regular.

<u>Remark.</u> Strictly speaking it is not necessary to consider an asymptotic structure in the image space defined in correspondence with 1.1. Only the <u>weak asymptotic structure</u> (see FELGENHAUER 1980a) defined by $\{O(E')\}$ for given expansion operators E' is needed. Instead of O_V we consider $O(E')$ and the question for regularity can be ignored. Instead of asymptotic continuity we then assume <u>weak asymptotic</u>

<u>continuity</u>

$$E'A(u+O_{\mathcal{U}}) = \{E'Au\} \quad . \quad \blacksquare$$

On condition that the solution of the singularly perturbed boundary value problem (1.23) has a boundary layer structure the used regular expansion operators E are usually composed of global and local expansion operators. Then for E' we take expansion operator composed analogously. We can restrict ourselves to composite operators of the form (1.8). Therefore let Z_i and Z'_i (i=1,...,k) be linear operators and let E_i, E'_i (i=1,...,k) be expansion operators in U and V. According to Theorem 1.2. the operators

$$E = \sum_{i=1}^{k} Z_i E_i \tag{1.28}$$

$$E' = \sum_{i=1}^{k} Z'_i E'_i \tag{1.29}$$

are composite expansion operators if and only if the matching-rules

$$\sum_i E_j Z_i E_i = E_j \ , \quad \sum_i E'_j Z'_i E'_i = E'_j \ , \qquad j = 1,\dots,k, \tag{1.30}$$

are satisfied. Close to (1.26) now we can formulate the approximation method

$$E'_i A u_i = E'_i f \ , \qquad u_i \in E_i \mathcal{U} \ , \qquad i = 1,\dots,k \ , \tag{1.31a}$$

$$\sum_i E_j Z_i u_i = u_j \ , \qquad\qquad j = 1,\dots,k \ . \tag{1.31b}$$

<u>Theorem 1.9.</u> Let the asymptotic methods (E_i, E'_i) be consistent at the point $u \in U$ for any i = 1,...,k and according to (1.28) and (1.29), respectively, let E, E' be the composite expansion operators. Then the asymptotic method (E,E') is consistent in u. ■

<u>Proof.</u> Because of the consistency from $E_i u = E_i E u$ it follows that

$E'_i A u = E'_i A E u$, consequently

$$E'Au = \sum_i Z'_i E'_i A u = \sum_i Z'_i E'_i A E u = E'AEu \ . \ \blacksquare$$

<u>Theorem 1.10.</u> Let the asymptotic methods (E_i, E'_i) be consistent. Then the approximation methods (1.26) and (1.31a), (1.31b) are equivalent, that means $\hat{u}_i = E_i \hat{u}$ is solution of (1.31) for every $\hat{u} \in \hat{U}$ and $\hat{u} = \sum Z_i \hat{u}_i$ satisfies the problem (1.26) for every set of solutions $\hat{u}_i$ of (1.31). ■

Proof. (i) $E_i' A E_i \hat{u} = E_i' A \hat{u} = E_i' E' A \hat{u} = E_i' E' f = E_i' f$

$$\sum_i E_j Z_i E_i \hat{u} = E_j \hat{u} \qquad \text{(because of (1.30))}$$

(ii) According to (1.31b) it follows $E_j \hat{u} = \hat{u}_j$ for $\hat{u} = \sum Z_i \hat{u}_i$. Consequently, we obtain

$$E' A \hat{u} = \sum Z_i' E_i' A \hat{u} = \sum Z_i' E_i' A E_i \hat{u} = \sum Z_i' E_i' A \hat{u}_i$$

$$= \sum Z_i' E_i' f = E' f \quad . \qquad \blacksquare$$

Now we consider an important special case. In many cases the operators E_i in U satisfy the simple matching-rule

$$E_i E_{i+1} = E_i \ , \qquad i = 1, \ldots, k-1 , \tag{1.32}$$

with the associated composite expansion operator $E = E_k$.
In this case it is useful to introduce the quantities $v_1 = E_1 u = u_1$, $v_i = (E_i - E_{i-1})u = u_i - u_{i-1}$ $(i=2,\ldots,k)$ for $u \in EU$. Then the problem (1.31) can be transformed into the approximation problem

$$E_1' A v_1 = E_1' f \ , \qquad v_1 \in E_1 U \ , \tag{1.33a}$$

$$\left.\begin{array}{l} E_i' A (v_i + \sum_{j=1}^{i-1} v_j) = E_i' f \\ E_{i-1} v_i = 0 \end{array}\right\} \quad v_i \in E_i U, \ i = 2, \ldots, k \ . \qquad \begin{array}{r}(1.33b)\\ (1.33c)\end{array}$$

This kind of a decomposition of the problem will be used at two different places. First, (1.31) is valid in the case of a finite asymptotic structure with more than one Landau-set (see(1.4)). Reducing the case $\omega = \{ O(\delta_\mu) \mid \mu = 1,\ldots,N\}$ to the case $\omega = \{ O(\delta_\mu) \mid \mu = N\}$ we do the same as if instead of the sequence of expansion operators $(E^\mu)_{\mu=0}^{N-1}$ only the operator E^{N-1} is considered as a composite expansion operator of all the E^μ. Then the problem (1.33) corresponds with the usual procedure of step by step expansion of the formal solution. The second application can be found in the method of the improvement of boundary layer. To explain this we consider the simplest case. Let E_1 be a global expansion operator and let $\tilde{E}$ be a local expansion operator appropriate to the problem under consideration. Further, let the matching-rule (1.11) be satisfied. Then the operator

$$E_2 = E_1 + \tilde{E} (I - E_1)$$

is a composite expansion operator of $\{E_1, \tilde{E}\}$ but also of $\{E_1, E_2\}$ and $E_1 E_2 = E_1$ is valid. Therefore, according to (1.33) we consider the approximation problem

$$E_1 A v_1 = E_1' f, \qquad v_1 \in E_1 \mathcal{U}$$
$$E_2' A (v_2 + v_1) = E_2' f, \quad v_2 \in E_2 \mathcal{U}$$
$$E_1 v_2 = 0 .$$

Hence we do not determine the local expansion $\tilde{E}\bar{u}$ of the formal solution but we define a <u>local corrector</u> $v_2 \in (E_2 - E_1)U = \tilde{E}(I-E_1)U = \tilde{E}O(E_1)$.

<u>1.4. Stability of asymptotic methods</u>

We have seen, that in the case of a consistent method with regular expansion operators asymptotic solutions are formal solutions. Up to now it has been an open question whether formal solutions are asymptotical. This main point can be considered as a problem of stability.

Furthermore another question is of importance, which has already been indicated in section 1.3. This is the question whether asymptotic solutions exist in the space considered. From the existence of solutions u_ε^* of the problem (1.23) for suffiently small ε it follows (without additional considerations) only that for the problem (1.25) there exists a solution in the space of all mappings $U_f = (\Lambda \to U_0)$ (as a rule u_ε^* is not continuous as a mapping of Λ in U_0 for singularly perturbed boundary value problems). However the usual expansion operators E composed of global and local ones are not regular in their maximal possible range of definition. Consequently, it is usual to suppose the existence of a linear subspace $U \subset U_f$ in which E is defined and regular such that $u^* = (u_\varepsilon^*)_{\varepsilon \in \Lambda} \in U$. Therefore, in addition to the construction of a formal solution it is necessary to prove that there exists an exact solution for which the formal solution is an asymptotic approximation.

Now we assume that U is chosen such that asymptotic solutions do exist, consequently $\tilde{U} \neq \emptyset$. In dependence on the construction of weak or strong formal solutions two concepts of stability are important.

<u>Definition 1.26.</u> Let A be an operator mapping a space $U = as_\Lambda(u_0, \delta)$ in a space $V = as_\Lambda(V_0, \delta')$ and let E' be an expansion operator in V. A is called <u>asymptotically stable</u> (<u>as-stable</u>) if for every $v \in U$ with $Av \in Au + O(\delta')$ it follows the relation $v \in u + O(\delta)$. It is called <u>E' -stable</u> at $u \in U$ if from $E'Av = E'Au$ it follows that $v \in u + O(\delta)$. ∎

<u>Remark.</u> In the case of a regular expansion operator E' both the concepts of stability are identical. ∎

<u>Theorem 1.11.</u> Let the set of exact solutions U^* be nonempty, let $\bar{u}$ be a strong formal solution of problem (1.25) and let A be as-stable at the point $\bar{u}$. Then $\bar{u}$ is an asymptotic solution. If $\hat{u}$ is a weak formal solution and A is E'-stable at the point $\hat{u}$, it follows that $\hat{u}$ is an

asymptotic solution. ■

<u>Proof.</u> Let u^* be an exact solution. With the as-stability from $A\bar{u} \in f + O(\delta')$ we obtain $\bar{u} \in u^* + O(\delta) \subset \tilde{U} = A^{-1}\{f\} + O(\delta)$.
In the same way, from $E'A\hat{u} = E'f$ it follows $\hat{u} \in u^* + O(\delta) \subset \tilde{U}$ by the E'-stability. ■

<u>Theorem 1.12.</u> Let $\bar{u}$ be a strong formal solution of (1.25) and let A be as-stable at $\bar{u}$. Then the strong formal solutions are asymptotically unique, that means $v \in \bar{u} + O(\delta)$ for every $v \in \bar{U}$. If $\hat{u}$ is weak formal solution and A is E'-stable at $\hat{u}$ every weak formal solution of (1.25) corresponds with $\hat{u}$. ■

<u>Proof.</u> In the case of an as-stable method if follows from $A\bar{u} \in f + O(\delta')$ and $Av \in f + O(\delta')$ that $\bar{u} \in v + O(\delta)$. For an E'-stable method, from $E'A\hat{u} = E'f = E'Av$ we obtain $\hat{u} \in v + O(\delta)$. Because of $v = Ev$, for every weak formal solution it follows that $\hat{u} = v$. ■

Let us indicate by two theorems how usual estimates can be used to produce statements of stability. Thereby we assume that the Landau-sets are defined by the relation (a) of Example 1.1. Analogous statements can also be found for other relations.

<u>Theorem 1.13.</u> We assume that there exist ε-independent constants $K_{\alpha\beta}$ and a real function $\gamma \in (\Lambda \to \mathbb{R})$ such that for every $\alpha \in \hat{\alpha}$ only a finite number of $K_{\alpha\beta}$ are non-zero and for the operator A with the seminorms of the basic spaces U_0, V_0 the a-priori estimates

$$p_\alpha(u-v) \leq \gamma(\varepsilon) \sum_\beta K_{\alpha\beta}\, q_\beta(Au-Av) \tag{1.34}$$

for every $\alpha \in \hat{\alpha}$, $u,v \in U$ are valid. If $\gamma\delta' < \delta$, then A is as-stable for every $u \in U$. ■

<u>Proof.</u> We note that from $q_\beta(Au-Av) < \delta'$ it follows that

$$p_\alpha(u-v) \leq \gamma \sum_\beta K_{\alpha\beta}\, q_\beta(Au-Av) \leq K\gamma\delta' < \delta . \quad ■$$

<u>Remark.</u> Applying this theorem we make use of the possibility that the seminorms can be considered to be ε-dependent. For instance in Theorem 2.9. we shall set $\hat{\alpha} = \{1\}$, $\hat{\beta} = \{1,2\}$, $p_1 = |\cdot|_{2m+\alpha}$,
$q_1 = \varepsilon^{-\frac{n}{2(2m-1)}}[\cdot]_0$, $q_2 = \varepsilon^{\frac{1+\alpha}{2m-1}}\|\cdot\|_\alpha$. Then we have $\gamma = 1$ and the operator associated with the problem (2.7) becomes as-stable with $\delta' = \delta$. ■

In many proofs of asymptotic including theorems are used. Therefore we shall consider partial orders $\leq_U$ in U and $\leq_V$ in V, respectively (which for instance follow by pointwise transformation of appropriate partial orders in U_0 and V_0, respectively for every $\varepsilon \in \Lambda$).

Definition 1.27. A partial order $\leq_{\mathcal{U}}$ in a space $U = as(U_0, \Delta)$ is called compatible with the asymptotic structure of the space if from

$$-v \leq_{\mathcal{U}} u \leq_{\mathcal{U}} v, \qquad v \in O(\delta)$$

it follows that $u \in O(\delta)$ for every $\delta \in \Delta$, $u, v \in U$. ■

In the case of maximum norms the usual pointwise defined partial order is compatible with the constructed asymptotic structure. A condition sufficient for compatibility is the following

Lemma 1.3. If from $-v \leq_{\mathcal{U}} u \leq_{\mathcal{U}} v$ it follows $p_\alpha(u) \leq p_\alpha(v)$ the partial order $\leq_{\mathcal{U}}$ is compatible with any asymptotic structure. ■

Proof. It follows from (1.1c). ■

Theorem 1.14. Let us assume that for the operator A an including theorem is valid, that means

$$Au \leq_V Av \quad \Longrightarrow \quad u \leq_{\mathcal{U}} v \quad . \tag{1.35}$$

Furthermore, let us assume that elements $\underline{s}$, $\bar{s} \in U$ called barrier functions can be found such that

$$A\underline{s} \leq_V f \leq_V A\bar{s} \quad , \tag{1.36a}$$

$$A\underline{s} \leq_V A\hat{u} \leq_V A\bar{s} \quad , \tag{1.36b}$$

$$\bar{s} - \underline{s} \in O(\delta) \quad . \tag{1.37}$$

If the partial order $\leq_{\mathcal{U}}$ is compatible with the asymptotic structure in U and if exact solutions $u^* \in U^*$ do exist, then the exact solutions are asymptotically unique and the relation

$$u^* \in \hat{u} + O(\delta) \quad , \tag{1.38}$$

is true for every $u^* \in U^*$. ■

Proof. According to the including theorem and (1.36) it follows $\underline{s} - \bar{s} \leq_{\mathcal{U}} \hat{u} - u^* \leq_{\mathcal{U}} \bar{s} - \underline{s}$. From (1.37) we obtain (1.38) and consequently $u_1^* \in u_2^* + O(\delta)$ for every u_1^*, $u_2^* \in U^*$. ■

Conclusion. Let the barrier functions in Theorem 1.14 be such that there exists for every $f \in V$ with $f \in O(\delta')$ a real number μ (independently of ε) such that

$$A(\mu\underline{s}) \leq_V f \leq_V A(\mu\bar{s}) \quad . \tag{1.39}$$

Then the operator A is as-stable. ■

Proof. From (1.37) we have $\mu\bar{s} - \mu\underline{s} \in O(\delta)$. ■

Remark. In the case of a linear operator the Theorems 1.13 and 1.14 can somewhat be simplified. Instead of (1.34) the estimates

$$p_\alpha(u) \leq \gamma \sum_\beta K_{\alpha\beta}\, q_\beta(Au)$$

are sufficient. Then an including theorem (1.35) is equivalent with a <u>maximum principle</u>

$$Au \leq_V 0 \qquad \Rightarrow \qquad u \leq_U 0 \quad .$$

As a rule it is possible then to suppose $-\underline{s} = \overline{s} = s$ and (1.39) transforms to

$$-\mu As \leq_V f \leq_V \mu As \quad . \qquad \blacksquare$$

1.5. Significant degeneration of differential operators

Now we consider the problem (1.23) that means for u, A and f the relations (1.24) are satisfied. For this purpose we use the special global and local expansion operators from the Definitions 1.13 and 1.14, respectively. Then the problems (1.31a) are the special global and local problems.

<u>Definition 1.28.</u> The problem

$$E^{\kappa}(\delta_0', \delta_1', \ldots, \delta_\kappa')(L_\varepsilon u)_{\varepsilon\in\Lambda} = E^{\kappa}(\delta_0', \delta_1', \ldots, \delta_\kappa') f$$
$$u = E^{\kappa}(\delta_0, \delta_1, \ldots, \delta_\kappa) u$$

is called the <u>global problem</u> of (1.23), the solutions of which are called <u>global formal solutions</u> of (1.23). ■

<u>Definition 1.29.</u> The problems

$$E^{\kappa}(\Gamma; \eta_1, \ldots, \eta_q; \delta_0', \ldots, \delta_\kappa')(L_\varepsilon u)_{\varepsilon\in\Lambda} = E^{\kappa}(\Gamma; \eta_1, \ldots, \eta_q; \delta_0', \ldots, \delta_\kappa') f$$
$$u = E^{\kappa}(\Gamma; \eta_1, \ldots, \eta_q; \delta_0', \ldots, \delta_\kappa') u$$

are called <u>local problems</u> (or boundary layer problems). Their solutions are called <u>local formal solutions</u>. ■ For a definite problem the $\delta_0, \ldots, \delta_\kappa$ and $\eta_1, \ldots, \eta_q$ are determined by hypothesises for the solution of (1.23). $\delta_0', \ldots, \delta_\kappa'$ have to be chosen such that the methods are consistent.

<u>Remark.</u> Global and local problems can be considered seperately as asymptotic methods. For the corresponding asymptotic structures Landau sets are defined on the basis of topologies of Example 1.2 (d) and (e), respectively. According to this asymptotic structures we call the asymptotic solutions <u>global solution</u> and <u>local solution</u> of (1.23), respectively. ■

From section 1.2 we know that global as well as local expansion operators can be defined by the formula

$$E^{\kappa} = E^{\kappa-1} + E_\kappa^0 (I - E^{\kappa-1})$$

for $k \geq 1$ (see Definition 1.13, 1.14). In both cases we have

$$E^{k-1} E^{k} = E^{k-1} \quad .$$

Under the condition of consistency of the global and local problems, respectively, the problems can be written in the form (see (1.33))

$$E_k^{o\prime} (L_\varepsilon (v_k + u_{k-1}) - L_\varepsilon u_{k-1}) = E^{k\prime} (f - L_\varepsilon u_{k-1}) \tag{1.40a}$$

$$v_k = E_k^{o} v_k \tag{1.40b}$$

where

$$u_k = u_{k-1} + v_k \quad .$$

<u>Remark.</u> From (1.40b) it follows that $E^{k-1} v_k = 0$. ∎

In the linear case this can be simplified to the form

$$E_k^{o\prime} L_\varepsilon v_k = E^{k\prime} (f - L_\varepsilon u_{k-1}) ,$$

$$v_k = E_k^{o} v_k \quad .$$

From this we see the importance of the operators $E_k^{o\prime} L_\varepsilon$ on the range of definition $E_k^{o} U$. $E_k^{o\prime} L_\varepsilon$ is called the corresponding <u>main part of L_ε</u> with respect to the transformation $(\eta_1, \dots, \eta_q)$.

From the above we obtain the following procedure for solving a linear singularly perturbed differential equation problem

$$L_\varepsilon u_\varepsilon \equiv \varepsilon L_1 u_\varepsilon + L_0 u_\varepsilon = f_\varepsilon \; , \qquad x \in \Omega \subset \mathbb{R}^n \; , \tag{1.41a}$$

$$l_\varepsilon u_\varepsilon = g_\varepsilon \; , \qquad x \in \Gamma \subset \partial\Omega \quad . \tag{1.41b}$$

1. Construction of a global formal solution

   Setting

$$u^N(x,\varepsilon) = \sum_{\mu=0}^{N} \varepsilon^\mu u_\mu(x)$$

   we obtain the problems

$$L_0 u_0 = f_0 \; , \qquad l_0 u_0 = g_0 \; ,$$

$$L_0 u_N = \lim_{\varepsilon \to 0} \varepsilon^{-N} (f_\varepsilon - \varepsilon L_1 u^{N-1}) \; , \qquad l_0 u_N = g_N \quad .$$

   As a rule the boundary conditions can be satisfied only on a part $\Gamma_- \subset \Gamma$ of the boundary. The global formal solution cannot always be determined independently of the construction of local correctors according to step 3 (see for instance section 4.3).

2. Hypothesis about the position of the boundary layers $\Gamma_i \subset \bar{\Omega}$.
In general boundary layers appear in a neighbourhood of points in which the solution u^N does not satisfy the boundary condition and at which the u_μ are discontinuous.

3. Construction of local formal solutions v_i^M.
Instead of the construction of local formal solutions we use so-called local correctors, that means instead of (1.41a,b) we start from

$$L_\varepsilon v_\varepsilon = f_\varepsilon - L_\varepsilon u^N - \sum_{j<i} L_\varepsilon v_{(j)}^{M_j} , \qquad (1.42a)$$

$$l_\varepsilon v_\varepsilon = g_\varepsilon - l_\varepsilon u^N - \sum_{j<i} l_\varepsilon v_{(j)}^{M_j} . \qquad (1.42b)$$

3.1. Introduction of local coordinates (ρ, φ).
Here $\rho = 0$ corresponds to the points of Γ_i.

3.2. Transformation $(\rho,\varphi) \to (\zeta,\varphi)$.
The equations

$$\zeta_r = \frac{\rho_r}{\varepsilon^{\beta_{r,i}}} , \qquad 1 \le r \le p ,$$

with appropriate $\beta_{r,i} > 0$ transform (1.42a,b) into

$$L_\varepsilon^{(i)} v_\varepsilon = L_0^{(i)} v_\varepsilon + M_\varepsilon^{(i)} v_\varepsilon = f_\varepsilon^{(i)} ,$$

$$v_\varepsilon |_{\zeta=0} = g^{(i)} .$$

In addition we have to satisfy conditions at infinity and, in general, boundary conditions with respect to φ.

3.3. Construction of local correctors
Setting

$$v_{(i)}^M(\zeta,\varphi,\varepsilon) = \sum_{\mu=0}^{M} \lambda^\mu v_\mu(\zeta,\varphi) , \qquad \varepsilon = \lambda^{\varkappa_i},$$

we obtain the problems

$$L_0^{(i)} v_0 = f_0^{(i)}$$

$$L_0^{(i)} v_M = \lim_{\varepsilon\to 0} \varepsilon^{-\frac{M}{\varkappa_i}} (f_\varepsilon^{(i)} - M_\varepsilon^{(i)} v_{(i)}^{M-1})$$

and additional conditions.

3.4. Extention of the range of definition on $\bar{\Omega}$.
As a rule the transformation to local coordinates is only possible in some neighbourhood of Γ_i. Therefore, we multiply the local correctors with appropriate cut off functions $\psi_i \in C^\infty(\bar{\Omega})$.

4. **Proof.**

It remains to show that the formal solution

$$\Phi_\varepsilon^N = u_N + \sum_i \psi_i v_{(i)}^{M_i}$$

is an asymptotic solution of (1.41a,b). Then in some norm the estimate

$$\| u_\varepsilon - \Phi_\varepsilon^N \| \leq K \varepsilon^{\varkappa}, \qquad \varkappa > 0 ,$$

is true.

It is an open problem to find the significant expansion operators from the set of all expansion operators $E^k(\Gamma; \eta_1, \ldots, \eta_q; \delta_0, \ldots, \delta_k)$. For many problems of the kind considered here the following way suggested by ECKHAUS is successful.

Analogously to section 1.2 expansion operators

$$\mathcal{E}^0(\Gamma; \eta_1, \ldots, \eta_q; \delta_0'; \delta_0)$$

can be defined in the set of all operators A of the considered kind. The functions $\eta_1, \ldots, \eta_q$ have the same meaning as in the definition of local expansion operators.

Thereby, $\mathcal{E}^0 A$ is determined in the following steps:

(i) We set $u_\varepsilon = \delta_0 v$ and formally transform the term $L_\varepsilon u_\varepsilon$ into a differential expression in the local coordinates (ζ, φ) (see definition 1.14).

(ii) We determine the formal limit

$$\hat{L}_0 v = \delta_0' \lim_{\varepsilon \to 0} \delta_0'^{-1} \hat{L}_\varepsilon v$$

where we interpret v and the derivatives of v with respect to (ζ, φ) as ε-independent.

(iii) We transform $\hat{L}_0 v$ to u_ε and to the original coordinates x.

<u>Definition 1.30.</u> $\mathcal{E}^0 A$ is called <u>degeneration of A</u>. $\mathcal{E}^0 A$ is called <u>significant degeneration</u> of A if $\mathcal{E}^0$, at the point A, is a significant expansion operator of a family of considered expansion operators for operators from $(U \to V)$. ∎

<u>Remark.</u> Restricting the studies to the subspace $E^0(\Gamma; \eta_1, \ldots, \eta_q; \delta_0)U$ we obtain

$$[\mathcal{E}^0(\Gamma; \eta_1, \ldots, \eta_q; \delta_0'; \delta_0) A] u = E^0(\Gamma; \eta_1, \ldots, \eta_q; \delta_0')(Au) \quad ,$$

that means the main part of the first approximation. ∎

Considering a family of transformations $(\eta_1, \ldots, \eta_q)$ for a boundary layer we can define expansion operators

$$E^k(\Gamma;\ \eta_1, \ldots, \eta_q;\ \delta_0, \ldots, \delta_k)$$

for the solutions of (1.23) and expansion operators

$$\epsilon^0(\Gamma;\ \eta_1, \ldots, \eta_q;\ \delta_0';\ \delta_0)$$

for the operator A. For the choice of used expansion operators E^k the following principle serves as hypothesis,

Correspondence principle. (ECKHAUS (1973), (1979)).

Let $E^k(\Gamma, \eta_1, \ldots, \eta_q;\ \delta_0, \ldots, \delta_k)$ be a significant expansion operator for the exact solution. Then $\epsilon^0(\Gamma;\ \eta_1, \ldots, \eta_q;\ \delta_0, \delta_0')$ A is a significant degeneration of operator A. ∎

It should be mentioned that the inversion of this hypothesis is, in general, incorrect.

Comments to Chapter 1

1.1. In analysis and, in particular, in mathematical physics it is a common practice to use LANDAU symbols. From the two possible exact interpretations of $f = O(g)$ (cf. BERG (1968)) as a relation between f and g or in the sense of set theory we have chosen the latter one. This makes it possible to reduce the convergence of asymptotic approximations and expansions to the usually applied convergence in topological spaces (FELGENHAUER (1980a)). The asymptotic approximations and expansions converge in the sense of a linear topology which is not locally convex (for the definition of linear topologies see KÖTHE (1960)). If we take into consideration the magnitude of the constant appearing in Example 1.1a, we come to locally convex topologies as given by PITTNAUER (1972).
Order functions have been investigated by KAPLUN et al. (KAPLUN (1967), LAGERSTROM (1972), see also ECKHAUS (1969), (1979)).
In the case that the basic spaces are BANACH spaces, conditions for the asymptotical completeness can be derived from the work of SWANSON/SCHULZER (1961).

1.2. ERDELYI (1956) was the first who referred to the fact that for the application of asymptotic methods it is necessary to introduce two kinds of definitions: asymptotic expansions in the sense of Definition 1.6 and asymptotic expansions generated by expansion operators (Definition 1.10). Expansion operators have been studied by FRAENKEL (1969) and in more detail, by ECKHAUS (1973), (1977a,b), (1979).

In connection with the matching rules in the Theorems 1.3 and 1.4 BERG (1976) was the first to interpret expansion operators as projectors in a linear space. By this more abstract consideration the global expansion operators have lost their excellent importance opposite the local ones. Therefore, in our opinion it is useful to understand, in difference to the general use, regularity of expansion operators in the sense of the Definitions 1.9 and 1.10 (for instance ECKHAUS (1973), (1979) means by regular expansion operators always the global ones - also if he studies singularly perturbed problems).
The notation "matching-rule" comes from the practice of the construction of asymptotic expansions. If a matching rule is satisfied (or if we suppose that it is satisfied), we obtain an additional rule for the determination of $E_1 u, \ldots, E_k u$, namely

$$M(E_1, \ldots, E_k)u = 0$$

(cf. formula 1.31b). A more detailed investigation concerning matching principles will be published in FELGENHAUER (1982), where also further matching rules will be given.
Basic works are FRAENKEL (1969), VAN DYKE (1975), ECKHAUS (1973), (1977b), MAUSS (1973), (1979). For the overlap hypothesis compare KAPLUN (1967), ECKHAUS (1969).

1.3. The chosen terminology follows from the possibility mentioned in the remarks to 1.1 to consider topologies for asymptotic methods. The formal solutions are then generalized solutions in the sense of these topologies which do not satisfy Hausdorff's axiom of separation. For different points of view concerning the asymptotic solutions of operator equations see SWANSON/SCHULZER (1961), BERG (1965), STENGLE (1971), BERGER/FRAENKEL (1971), LIONS (1973).

1.4. In general, all estimates which yield statements about stability can be used for the proof of asymptotical stability. More general statements can be found in ECKHAUS (1977), (1979) and REINHARD (1978) and in the bibliography given in point 1.3. See also Chapter 2.

1.5. The correspondence principle has been formulated by ECKHAUS (1973), (1979). There have also been given counter examples. An exact foundation connected with assumptions for the validity is still open.

2. ESTIMATES OF THE SOLUTIONS OF ELLIPTIC AND PARABOLIC BOUNDARY VALUE PROBLEMS

In this chapter we shall consider elliptic and parabolic boundary value problems of the form

$$L_\varepsilon u_\varepsilon \equiv \varepsilon L_1 u_\varepsilon + L_0 u_\varepsilon = f_\varepsilon \qquad \text{in } \Omega$$
$$\frac{\partial^s u_\varepsilon}{\partial \nu^s} = 0 \qquad \text{on } \partial\Omega, \quad s = 0,1,\dots,m-1 \tag{2.1$_e$}$$

and

$$\frac{\partial u_\varepsilon}{\partial t} + L_\varepsilon u_\varepsilon = f_\varepsilon \qquad \text{in } Q$$
$$\frac{\partial^s u_\varepsilon}{\partial \nu^s} = 0 \qquad \text{on } S_T, \quad s = 0,1,\dots,m-1, \tag{2.1$_p$}$$
$$u_\varepsilon = 0 \qquad \text{in } \Omega$$

respectively where Ω denotes a bounded domain of $\mathbb{R}^n$ with sufficiently regular boundary $\partial\Omega$, $Q = \Omega \times (0,T]$, $S_T = \partial\Omega \times [0,T]$ and ν is the outward directed unit normal vector. In the case of second order problems (m=1) we also consider special domains with non-regular boundary. Let L_1 be a linear, uniformly elliptic differential operator of order 2m

$$L_1 = \sum_{|\beta| \le 2m} a_\beta D^\beta$$

and let L_0 be a linear differential operator of order 1

$$L_0 = \sum_{\mu=1}^{n} b_\mu \frac{\partial}{\partial x_\mu} + c \qquad .$$

The coefficients of L_1, L_0 and f_ε are assumed to be sufficiently regular on $\bar{\Omega}$ and $\bar{Q}$, respectively. The aim of this chapter is to clarify the existence and estimates of classical solutions of $(2.1)_e$ and $(2.1)_p$. In section 2.5 we shall give a method for studying nonlinear problems. Thereby estimates of derivatives of solutions of linear problems with ε-dependent coefficients are useful.

2.1. Notations and Definitions

For sufficiently regular functions u we denote

$$[u]_\mu = \sum_{|\beta|=\mu} \sup_{x\in\Omega} |D^\beta u(x)| , \qquad \mu \ge 0 \text{ integer}$$

$$[u]_{\mu+\alpha} = \sum_{|\beta|=\mu} \sup_{\substack{x,y\in\Omega \\ x\neq y}} \frac{|D^\beta u(x) - D^\beta u(y)|}{|x-y|^\alpha} , \quad 0<\alpha<1 , \ \mu \ge 0, \text{integer}$$

$$\langle u \rangle = \sup_{(x,t)\in Q} |u(x,t)|$$

$$\langle u\rangle_{\alpha,x} = \sup_{\substack{(x,t),(y,t)\in Q\\ x\neq y}} \frac{|u(x,t)-u(y,t)|}{|x-y|^{\alpha}}\ , \quad 0<\alpha<1\ ,$$

$$\langle u\rangle_{\alpha,t} = \sup_{\substack{(x,t),(x,\tau)\in Q\\ t\neq\tau}} \frac{|u(x,t)-u(x,\tau)|}{|t-\tau|^{\alpha}}\ , \quad 0<\alpha<1\ .$$

If it is clear from the context whether it is the question of functions defined on Ω or $Q = \Omega \times (0,T]$ we shall also briefly write

$$\|u\| = [u]_0 \qquad \text{or} \qquad \|u\| = \langle u\rangle\ .$$

We shall look for classical solutions in the Banach-spaces $C^{2m+\alpha}(\bar{\Omega})$ and $C^{2m+\alpha,\frac{2m+\alpha}{2m}}(\bar{Q})$, respectively. The norms in $C^{l+\alpha}(\bar{\Omega})$ and $C^{l+\alpha,\frac{l+\alpha}{2m}}(\bar{Q})$, respectively, for $l = 0,1,\ldots$ are given by

$$\|u\|_{l+\alpha} = \sum_{\mu=0}^{l} ([u]_{\mu} + [u]_{\mu+\alpha})$$

$$\|u\|_{l+\alpha,2m} = \sum_{|\beta|+2m\gamma\le l} \langle D_x^{\beta} D_t^{\gamma} u\rangle + \sum_{|\beta|+2m\gamma=l} \langle D_x^{\beta} D_t^{\gamma} u\rangle_{\alpha,x} +$$

$$+ \sum_{0<l-2m\gamma-|\beta|<2m} \langle D_x^{\beta} D_t^{\gamma} u\rangle_{\frac{l+\alpha-2m\gamma-|\beta|}{2m},t}\ , \qquad m = 1,2,\ldots\ .$$

In the space $C^{2m+\alpha}(\bar{\Omega})$ we furthermore introduce the ε-dependent norm

$$|u|_{2m+\alpha} = \sum_{\mu=0}^{2m} (\varepsilon^{\frac{\mu}{2m-1}}[u]_{\mu} + \varepsilon^{\frac{\mu+\alpha}{2m-1}}[u]_{\mu+\alpha})$$

which is finer than the maximumnorm in $C(\bar{\Omega})$ and is equivalent to $\|\cdot\|_{2m+\alpha}$ for every fixed $\varepsilon > 0$.

2.2. Existence and a-priori-estimates of classical solutions

We consider the elliptic boundary value problem

$$Lu \equiv \sum_{|\beta| \le 2m} a_\beta(x)\, D^\beta u = f(x) \qquad \text{in } \Omega$$
$$\frac{\partial^s u}{\partial \nu^s} = 0 \qquad \text{on } \partial\Omega,\ s = 0,\dots,m-1 \tag{2.2}$$

under the conditions

(H1) L is <u>uniformly elliptic</u>, that means there is a constant $\alpha_0 > 0$ such that for every $y \in \mathbb{R}^n$, $x \in \Omega$ the inequality

$$(-1)^m \sum_{|\beta| = 2m} a_\beta(x)\, y^\beta \ge \alpha_0 \Big(\sum_{\mu=1}^{n} y_\mu^2 \Big)^m$$

is satisfied.

(H2) For some $l \ge 2m$ and some α with $0 < \alpha < 1$ let $a_\beta \in C^{l-2m+\alpha}(\bar\Omega)$, $\partial\Omega \in C^{l+\alpha}$.

Then we have

<u>Theorem 2.1.</u> (AGMON, DOUGLIS, NIRENBERG (1959))
Assuming (H1) and (H2), the a-priori estimate

$$\| u \|_{l+\alpha} \le K \big(\| Lu \|_{l-2m+\alpha} + [u]_0 \big)$$

is valid for every $u \in C^{l+\alpha}(\bar\Omega)$ with $\frac{\partial^s u}{\partial \nu^s} = 0$ on $\partial\Omega$, $s = 0,1,\dots,m-1$. If there exists at most one solution of (2.2) in $C^{l+\alpha}(\bar\Omega)$, then there is exactly one solution $u \in C^{l+\alpha}(\bar\Omega)$ of (2.2) for every $f \in C^{l-2m+\alpha}(\bar\Omega)$. ∎

We now consider the parabolic boundary value problem

$$\frac{\partial u}{\partial t} + Lu \equiv \frac{\partial u}{\partial t} + \sum_{|\beta| \le 2m} a_\beta(x,t)\, D^\beta u = f(x,t) \qquad \text{in } Q$$
$$\frac{\partial^s u}{\partial \nu^s} = 0 \qquad \text{on } S_T \tag{2.3}$$
$$u = 0 \qquad \text{in } \Omega$$

under the conditions

(H1') There is a constant $\alpha_0 > 0$, such that for every $y \in \mathbb{R}$, $(x,t) \in \bar Q$ the inequality

$$(-1)^m \sum_{|\beta| = 2m} a_\beta(x,t)\, y^\beta \ge \alpha_0 \Big(\sum_{\mu=1}^{n} y_\mu^2 \Big)^m$$

is satisfied. Then $\frac{\partial}{\partial t} + L$ is called <u>uniformly parabolic.</u>

(H2') For some $l \geq 2m$ and some α with $0 < \alpha < 1$,

let a_β, $f \in C^{l-2m+\alpha, \frac{l-2m+\alpha}{2m}}(\bar{Q})$, $\partial\Omega \in C^{l+\alpha}$.

(H3) Let the <u>compatibility conditions</u> of order l be fulfilled, that means

$$\left.\frac{\partial^i f(x,t)}{\partial t^i}\right|_{t=0} \equiv 0 , \qquad i = 0,\dots,\left[\frac{l-m+1}{2m}\right] .$$

Then it is true the following

<u>Theorem 2.2.</u> (SOLONNIKOV (1965))

Assuming (H1'), (H2') and (H3) there is a unique solution $u \in C^{l+\alpha, \frac{l+\alpha}{2m}}(\bar{Q})$ of (2.3) for every $f \in C^{l-2m+\alpha, \frac{l-2m+\alpha}{2m}}(\bar{Q})$ and the a-priori-estimate

$$\| u \|_{l+\alpha, 2m} \leq K \| f \|_{l-2m+\alpha, 2m}$$

is valid. ∎

<u>Remark.</u> Concerning statements on existence it is possible to weaken the hypothesis(H3).However, for the construction of asymptotic approximations in the sections 4.2 and 4.3 we shall always suppose compatibility of order $\tilde{l}$ with sufficiently large $\tilde{l}$. ∎

In the sections 4.4 and 4.5 we shall discuss boundary value problems in non-regular bounded domains for linear, uniformly elliptic problems of second order.

Let

$$Lu \equiv \sum_{|\beta| \leq 2} a_\beta(x) D^\beta u = f(x) \qquad \text{in } \Omega$$
$$u = g \qquad \text{on } \partial\Omega \tag{2.4}$$

be given and let $\partial\Omega \notin C^{2+\alpha}$ for every α with $0 < \alpha < 1$. To guarantee that classical solutions of (2.4) are elements of the space $C^{2+\alpha}(\bar{\Omega})$ it is necessary to satisfy corresponding compatibility conditions (AZZAM (1980)). However, in $C(\bar{\Omega}) \cap C^2(\Omega)$ the unique classical solvability can be guaranteed for a large class of bounded domains. To characterize the admissible domains we define analoguous to (MICHAEL (1977)).

<u>Definition 2.1</u> We say that there exists a (p,P)-barrier for the bounded domain Ω, if there is a function $\chi \in C(\bar{\Omega}) \cap C^2(\Omega)$ with the following properties

(i) $\chi = 0$ on $\partial\Omega$, $\chi > 0$ in Ω

(ii) there are positive constants m_1, m_2 such that for $\rho(x) = \operatorname{dist}(x, \partial\Omega)$ the inequalities

$$\chi(x) \geq m_1 (\rho(x))^2, \qquad |\nabla \chi(x)| \leq m_2 \frac{\chi(x)}{\rho(x)} \qquad in\ \Omega$$

are valid

(iii) there is a positive constant m_3 such that if a_{ij} is a symmetric matrix with $|a_{ij}| \leq P$ and $\sum_{i,j=1}^{n} a_{ij} y_i y_j \geq p \sum_{i=1}^{n} y_i^2$,

the inequality

$$\sum_{i,j=1}^{n} a_{ij} \frac{\partial^2 \chi}{\partial x_i \partial x_j}(x) \leq -m_3 \chi(x) (\rho(x))^{-2}$$

is satisfied for all $x \in \Omega$.

A bounded domain Ω is called an admissible domain if for all (p,P) with $0 < p \leq P$ a (p,P)-barrier for Ω does exist. ■

Examples for admissible bounded domains are $\prod_{i=1}^{n} (y_i, z_i)$, open balls in $\mathbb{R}^n$ and bounded domains with regular boundary. Furthermore, the nonempty intersection of a finite number of admissible bounded domains is also admissible.

We now weaken (H2) to

(H4) Let Ω be an admissible bounded domain, let a_β, f be local Hölder continuous with exponent α with $0 < \alpha < 1$ and let g be continuous on $\partial\Omega$. Furthermore, let positive constants K_1, K_2 with $K_1 \leq 1$ exist , such that for all $x \in \Omega$ the estimates

$$|a_\beta(x)| \leq K_2 \ for\ |\beta| = 2, \qquad |a_\beta(x)| \leq K_2 (\rho(x))^{K_1 - 1} \ for\ |\beta| = 1$$

$$|a_0(x)| \leq K_2 (\rho(x))^{K_1 - 2}, \ |f(x)| \leq K_2 (\rho(x))^{K_1 - 2} \quad and \quad a_0(x) \geq 0$$

are satisfied.

Then it follows

Theorem 2.3. (MICHAEL (1977))
Assuming (H1), (H4) there exists a classical solution of (2.4) which is unique and continuous on $\bar{\Omega}$ and belongs to $C^{2+\alpha}(M)$ for any compact subset M of Ω . ■

To estimate classical solutions of (2.4) we use a majoration variant of the maximum principle.

Theorem 2.4. (PROTTER/WEINBERGER (1967))
Let L be uniformly elliptic, let a_β be bounded and let Ω be a bounded domain. If for two functions u,S belonging to $C(\bar{\Omega}) \cap C^2(\Omega)$ the inequalities

$|(Lu)(x)| \le (LS)(x)$ for every $x \in \Omega$, $|u(x)| \le S(x)$ for every $x \in \partial\Omega$

are satisfied, then for every $x \in \bar{\Omega}$ the estimate

$$|u(x)| \le S(x)$$

is valid. ∎

In the following the function S majorizing u is called a <u>barrier function</u>.

We now consider parabolic boundary value problems of second order in the form

$$\frac{\partial u}{\partial t} + Lu \equiv \frac{\partial u}{\partial t} - \sum_{|\beta| \le 2} a_\beta(x,t)\, D^\beta u = f(x,t) \qquad \text{in } Q \tag{2.5}$$
$$u = g \qquad \text{on } S_T \cup (\Omega \times \{0\})$$

and do not necessarily suppose compatibility (H3) of the data. As a rule the derivatives of classical solutions of (2.5) are no longer uniformly continuous on Q. Therefore we shall study the solvability in the spaces $C_{2+\alpha}$ (FRIEDMAN (1964)). (H2) and (H3) are weakened to

(H5) Let g be continuous on $S_T \cup (\Omega \times \{0\})$, let a_β, f be Hölder continuous with the exponent α with $0 < \alpha < 1$ and let their C_α-norms be bounded. Suppose that Ω is such that for every $\tilde{x} \in \partial\Omega$ there is a ball B with the centre $\bar{x}$ such that $B \cap \bar{\Omega} = \{\tilde{x}\}$ and $|\bar{x} - x| \ge \mu(x) > 0$ for every $x \in \bar{\Omega}$.

Theorem 2.5. (FRIEDMAN (1964))
Assuming (H1'), (H5), there is exactly one classical solution of (2.5) belonging to $C_{2+\alpha}$. ∎

Remark. From $u \in C_{2+\alpha}$ it follows in particular that u is continuous on $\bar{Q}$. ∎

Analogously to the elliptic case we use a majoration variant of the maximum principle to estimate classical solutions of (2.5).

Theorem 2.6. (PROTTER/WEIBERGER (1967))
Let $\frac{\partial}{\partial t} + L$ be uniformly parabolic, let a_β be bounded, let Ω be a bounded domain, $Q = \Omega \times (0,T]$ and let u, S be continuous functions on $\bar{Q}$, being continuously differentiable in Q, twice with respect to x and once with respect to t. Then from

$$|(\frac{\partial u}{\partial t} + Lu)(x,t)| \leq (\frac{\partial S}{\partial t} + LS)(x,t) \qquad \text{in } Q$$

$$|u(x,t)| \leq S(x,t) \qquad \text{on } S_T \cup (\Omega \times \{0\})$$

it follows the estimate

$$|u(x,t)| \leq S(x,t) \qquad \text{in } \bar{Q} \quad . \; \blacksquare$$

Remark. Theorem 2.6 is also true for noncylindrical domains Q of the structure $Q = \{(x,t) \mid x \in \Omega_t , \; 0 < t \leq T\}$, where Ω_t is a bounded domain of $\mathbb{R}^n$. This will be applied in section 4.4. ■

It is not always possible to satisfy the regularity conditions for u and S. Using the generalized maximum principle, no additional technical difficulties arise.

Let $H^1_0(\Omega)$ be the completion in the norm

$$\|v\|^2_{H^1(\Omega)} = \sum_{|\beta| \leq 1} \int_\Omega |D^\beta v(x)|^2 dx$$

of the set of all functions infinitely differentiable in Ω which have a compact support, let $H^1(\Omega)$ be the set of all functions whose generalized derivatives of orders $|\beta| \leq 1$ belong to $L^2(\Omega)$ and let $H^{-1}(\Omega)$ be the dual space with respect to $H^1_0(\Omega)$. The scalar products on $L^2(\Omega) \times L^2(\Omega)$ and $H^{-1}(\Omega) \times H^1(\Omega)$, respectively, are denoted by $(\cdot\, , \cdot)$ and $\langle \cdot\, , \cdot \rangle$. $v \geq 0$ in H^1 means that v is pointwise almost everywhere nonnegative. The inequality $w \geq 0$ in H^{-1} is equivalent to $\langle w, v\rangle \geq 0$ for all $v \in H^1_0$ with $v \geq 0$. The bilinear form

$$\langle L(t)u,v\rangle = \int_\Omega \Big(\sum_{i,j=1}^n a_{ij}(x,t) \frac{\partial u}{\partial x_j} \frac{\partial v}{\partial x_i} + \sum_{j=1}^n b_j(x,t) \frac{\partial u}{\partial x_j} v + c(x,t)uv \Big) dx$$

defined on $H^1(\Omega) \times H^1(\Omega)$ is related to the classical differential operator

$$(Lv)(x,t) \equiv -\sum_{i,j=1}^n \frac{\partial}{\partial x_i}\Big(a_{ij}(x,t) \frac{\partial v}{\partial x_j}\Big) + \sum_{j=1}^n b_j(x,t) \frac{\partial v}{\partial x_j} + c(x,t)v \; . \qquad (2.6)$$

Finally, by u(t) we denote the element $u(\cdot,t)$ of $H^1(\Omega)$.

We assume

(H6) for almost every $t \in (0,T)$ $L(t)$ is monotonic on $H^1(\Omega)$ that means

$\langle L(t)v,v\rangle \geq 0$ for all $v \in H^1_0(\Omega)$ and almost all $t \in (0,T)$.

Theorem 2.7. Assume that $u \in H^1(Q)$ and that (H6) is fulfilled. Then the inequalities

$$u'(t) + L(t)u(t) \leq 0 \qquad a.e. \text{ in } (0,T)$$

$$u(0) \leq 0$$

$$u(x,t) \leq 0 \qquad \text{for } x \in \partial\Omega, \; a.e. \text{ in } (0,T)$$

imply the estimate

$$u(x,t) \leq 0 \qquad \text{a.e. in } Q \ . \quad \blacksquare$$

Remarks (i) The assumption $u \in H^1(Q)$ guarantees the continuity of u as a mapping from $[0,T]$ into $L^2(\Omega)$. Thus u(o) makes sense.

$$u(x,t) \leq 0 \ , \qquad x \in \partial\Omega$$

is equivalent to the fact that there exists a $v \in H_0^1(\Omega)$ with $v - u(t) \geq 0$.

(ii) the condition

$$c(x,t) - \frac{1}{2}\sum_{\mu=1}^{n} \frac{\partial b_\mu}{\partial x_\mu}(x,t) \geq 0$$

is sufficient for (H6). $\blacksquare$

Proof of Theorem 2.7. Setting $u^+(x,t) = \max(u(x,t), 0)$ we obtain $u^+ \in H^1(Q)$,

$$\frac{\partial u^+}{\partial t}(x,t) = \begin{cases} \frac{\partial u}{\partial t}(x,t) & \text{a.e. in } Q \text{ where } u(x,t) > 0 \\ 0 & \text{a.e. in } Q \text{ where } u(x,t) \leq 0 \end{cases}$$

$$\frac{\partial u^+}{\partial x_\mu}(x,t) = \begin{cases} \frac{\partial u}{\partial x_\mu}(x,t) & \text{a.e. in } Q \text{ where } u(x,t) > 0 \\ 0 & \text{a.e. in } Q \text{ where } u(x,t) \leq 0 \end{cases}$$

(cf. e.g. LADYZHENSKAYA/URALTSEVA (1973)). Then for fixed $t \in (0,T)$ the set $\Omega(t) = \{x \in \Omega \mid u(x,t) > 0\}$ is measurable and we have

$$\langle u'(t), u^+(t)\rangle = \int_{\Omega(t)} \frac{\partial u}{\partial t}(x,t)\, u^+(x,t)\,dx = \int_{\Omega(t)} \frac{\partial u^+}{\partial t}(x,t)\, u^+(x,t)\,dx$$

$$= \langle u^{+\prime}(t), u^+(t)\rangle$$

and analogously

$$\langle L(t)u(t), u^+(t)\rangle = \int_{\Omega(t)} \Big(\sum_{i,j=1}^{n} a_{ij}(x,t)\frac{\partial u}{\partial x_j}\frac{\partial u^+}{\partial x_i} + \sum_{j=1}^{n} b_j(x,t)\frac{\partial u}{\partial x_j}u^+ + c(x,t)\,u\,u^+\Big)dx$$

$$= \langle L(t)u^+(t), u^+(t)\rangle \geq 0 \quad .$$

The assumptions imply that

$$\langle u'(t), v\rangle + \langle L(t)u(t), v\rangle \leq 0 \quad \text{for all } v \in H_0^1(\Omega) \text{ with } v \geq 0, \text{a.e. in } (0,T).$$

Choosing $v = u^+(t) \in H_0^1(\Omega)$, from the above, by integration over t we obtain

$$\frac{1}{2}[(u^+(t),u^+(t)) - (u^+(0),u^+(0))] = \int_0^t \langle u^{+\prime}(t), u^+(t)\rangle dt \leq 0, \quad t \in (0,T).$$

From the initial condition it finally follows that $u^+(0) = 0$ such that

$$u^+(x,t) = 0 \qquad a.e. \text{ in } Q$$

which implies the assertion. ■

Theorem 2.7 can be considered as a starting-point for a majoration variant analogous to Theorem 2.6. In section 4.1 it will be applied to give estimates of the global asymptotic solution under weaker assumptions on u and S.

2.3. A-priori estimates for singularly perturbed elliptic problems

In the application of the a-priori estimates given in the Theorems 2.1 and 2.2 to the solution of singularly perturbed elliptic and parabolic problems, respectively of the form $(2.1)_e$, $(2.1)_p$ the constant which appeared in these estimates depends on the parameter ε in an unknown way. Both this and the next section aim at giving a-priori estimates for which the dependence of ε is explicitly known.

We first consider the elliptic boundary value problem

$$\begin{aligned} L_\varepsilon u_\varepsilon \equiv \varepsilon L_1 u_\varepsilon + L_0 u_\varepsilon = f_\varepsilon \qquad & \text{in } \Omega \\ \frac{\partial^s u_\varepsilon}{\partial \nu^s} = 0 \qquad & \text{on } \partial\Omega, \quad s = 0,1,\dots,m-1 \end{aligned} \tag{2.7}$$

with

$$L_1 \equiv \sum_{|\beta| \leq 2m} a_\beta(x) D^\beta, \qquad L_0 \equiv \sum_{\mu=1}^{n} b_\mu(x) \frac{\partial}{\partial x_\mu} + c(x),$$

in a bounded domain Ω with sufficiently regular boundary $\partial\Omega$. Let L_1 be uniformly elliptic and let a_β, b_μ, c as well as f_ε be sufficiently regular. Then it follows

Theorem 2.8. For any classical solution $u \in C^{2m+\alpha}(\bar{\Omega})$ of (2.7) and for sufficiently small ε the a-priori estimate

$$|u|_{2m+\alpha} \leq K\left(\varepsilon^{\frac{1+\alpha}{2m-1}} \| L_\varepsilon u\|_\alpha + [u]_0\right)$$

is valid with a constant K being independent of ε. ■

For the proof of Theorem 2.8 the following interpolation inequality is needed.

Lemma 2.1. (MIRANDA (1970)) For all real numbers p, q with $0 \le p < q$ there is a constant K only dependent on p, q and Ω such that for $u \in C^q(\bar{\Omega})$ and all $\omega \in (0, \omega_0)$

$$[u]_p \le \omega [u]_q + K \omega^{-\frac{p}{q-p}} [u]_0$$

is valid. ■

Proof of Theorem 2.8. Applying Theorem 2.1 to the boundary value problem

$$L_1 u = \frac{1}{\varepsilon}(L_\varepsilon u - L_0 u) \qquad \text{in } \Omega$$

$$\frac{\partial^s u}{\partial \nu^s} = 0 \qquad \text{on } \partial\Omega,\ s = 0,1,\dots,m-1$$

we obtain

$$[u]_{2m+\alpha} \le \|u\|_{2m+\alpha} \le \frac{K}{\varepsilon} \|L_\varepsilon u\|_\alpha + \frac{K}{\varepsilon} \|L_0 u\|_\alpha + K[u]_0 .$$

From Lemma 2.1 it follows that

$$\|L_0 u\|_\alpha \le K\left([u]_{1+\alpha} + [u]_0\right)$$

and

$$\frac{K}{\varepsilon} \|L_0 u\|_\alpha \le \frac{K}{\varepsilon} \omega [u]_{2m+\alpha} + \frac{K}{\varepsilon} \omega^{-\frac{1+\alpha}{2m-1}} [u]_0 + \frac{K}{\varepsilon} [u]_0 .$$

Choosing $\omega = \frac{\varepsilon}{2K}$ we obtain, for sufficiently small ε, that

$$[u]_{2m+\alpha} \le \frac{K}{\varepsilon} \|L_\varepsilon u\|_\alpha + K \varepsilon^{-\frac{2m+\alpha}{2m-1}} [u]_0 .$$

Taking into consideration the definition of $|\cdot|_{2m+\alpha}$ from Lemma 2.1 it follows the assertion of Theorem 2.8 for appropriate choice of ω . ■

In the case m = 1 we shall eliminate $[u]_0$ by applying the maximum principle. For m > 1 we estimate $[u]_0$ by the norm of Hilbert-space $L^2(\Omega)$ of u and use the semiboundedness of L_1 in $L^2(\Omega)$ and the special form of L_0.

Lemma 2.2. (BESJES (1975))
Let Ω be a bounded domain with regular boundary. Then there are positive constants K and ω_0 such that for all $\omega \in (0, \omega_0)$ and $u \in C^1(\bar{\Omega})$

$$[u]_0 \le K \omega^{-\frac{n}{2}} \left(\int_\Omega |u(x)|^2 dx \right)^{\frac{1}{2}} + \omega [u]_1$$

is valid. ■

Theorem 2.9. Let L_1 be a linear, uniformly elliptic differential operator of order 2m, $L_0 \equiv \sum_{\mu=1}^{n} b_\mu(x) \frac{\partial}{\partial x_\mu} + c(x)$ with $c(x) - \frac{1}{2}\operatorname{div} b(x) \geq K_0 > 0$ and let $\partial\Omega$ as well as the coefficients of L_1, L_0 and f_ε be sufficiently regular. Then, for all sufficiently small ε there exists exactly one solution u_ε of (2.7) satisfying the estimate

$$|u_\varepsilon|_{2m+\alpha} \leq K\left(\varepsilon^{-\frac{n}{2(2m-1)}}[f_\varepsilon]_0 + \varepsilon^{\frac{1+\alpha}{2m-1}}\|f_\varepsilon\|_\alpha\right)$$

with a constant K being independent of ε. ∎

Proof. From Theorem 2.8 we first of all obtain

$$\varepsilon^{\frac{1}{2m-1}}[u_\varepsilon]_1 \leq K\left(\varepsilon^{\frac{1+\alpha}{2m-1}}\|f_\varepsilon\|_\alpha + [u_\varepsilon]_0\right) .$$

Setting $\omega = \frac{1}{2K}\varepsilon^{\frac{1}{2m-1}}$ in Lemma 2.2 we can eliminate $[u_\varepsilon]_1$. For sufficiently small ε it follows that

$$[u_\varepsilon]_0 \leq K\left\{\varepsilon^{\frac{1+\alpha}{2m-1}}\|f_\varepsilon\|_\alpha + \varepsilon^{-\frac{n}{2(2m-1)}}\left(\int_\Omega |u_\varepsilon(x)|^2 dx\right)^{\frac{1}{2}}\right\} . \qquad (2.8)$$

From Garding's inequality we have the existence of a positive constant K such that

$$\int_\Omega u(x)(L_1 u)(x)\,dx \geq -K\int_\Omega |u(x)|^2 dx .$$

Consequently,

$$\int_\Omega u(x)(\varepsilon L_1 u + L_0 u)(x)\,dx \geq \int_\Omega \left(c(x) - \tfrac{1}{2}\operatorname{div} b(x) - \varepsilon K\right)|u(x)|^2 dx$$

and for sufficiently small ε

$$c(x) - \tfrac{1}{2}\operatorname{div} b(x) - \varepsilon K \geq \frac{K_0}{2} > 0 .$$

Using the inequality of Cauchy-Schwarz, (2.8) and Theorem 2.8 we obtain the a-priori estimates of Theorem 2.9 being valid for any classical solution of (2.7). However, there exists at most one classical solution and so the existence of a unique classical solution is guaranteed according to Theorem 2.1. ∎

Now we consider the elimination of $[u]_0$ in the estimate of Theorem 2.8 in the case of elliptic problems of second order. For this purpose we use the maximum principle (see Theorem 2.4) and construct special adapted barrier functions in dependence on the concrete form of L_0.

(i) the case $c(x) \geq c_0 > 0$.

Then $S_\varepsilon(x) = \frac{2}{c_0} [f_\varepsilon]_0$ is a function majorizing u_ε and it follows

$$[u_\varepsilon]_0 \leq \frac{2}{c_0} [f_\varepsilon]_0 \ . \tag{2.9}$$

(ii) the case $b_n(x) \neq 0$.

Without restriction of generality let $b_n(x) \geq b_0 > 0$. Because of boundedness of Ω there are two numbers $\gamma_1, \gamma_2 \in \mathbb{R}$ with $\gamma_1 \leq x_n \leq \gamma_2$ in $\overline{\Omega}$. Then

$$S_\varepsilon(x) \equiv \frac{2}{b_0} [f_\varepsilon]_0 (x_n - \gamma_1)$$

is a function majorizing u_ε and it follows

$$[u_\varepsilon]_0 \leq \frac{2}{b_0} (\gamma_2 - \gamma_1) [f_\varepsilon]_0 \ . \tag{2.10}$$

The estimates (2.9) and (2.10) are special cases of a result of VAN HARTEN (1978):

If there are constants $\gamma_0, \gamma_1, \ldots, \gamma_n$ such that

$$\sum_{\mu=1}^{n} \gamma_\mu b_\mu(x) + c(x) \geq \gamma_0 > 0 \qquad in\ \overline{\Omega}, \tag{2.11}$$

then for sufficiently small ε it follows

$$[u_\varepsilon]_0 \leq K [f_\varepsilon]_0$$

with a constant K being independent of ε.

We remark that (2.11) is not satisfied if b_μ, c commonly vanish in $\overline{\Omega}$ (turning point case).

(iii) the case $b_n(x) = x_n^l\, b^*(x)$, $b^*(x) \neq 0$, l odd, $l \geq 1$

Let first of all $b^*(x) \leq -b_0 < 0$ and $l = 2m - 1$. The function

$$w_\varepsilon(t) = \int_0^{t\varepsilon^{\frac{-1}{2m}}} \int_0^{q} exp\, \frac{A}{2m} (p^{2m} - q^{2m})\, dp\, dq$$

has the following properties for every $A > 0$ and for sufficiently small ε

P1: $\varepsilon w_\varepsilon''(t) + A t^{2m-1} w_\varepsilon'(t) = \varepsilon^{1-\frac{1}{m}}$

P2: For all $t \in [t_1, t_2]$ it is valid that

$$|w_\varepsilon(t)| \leq K\, |ln\, \varepsilon| \quad for\ m = 1, \qquad |w_\varepsilon(t)| \leq K \quad for\ m > 1$$

and

$$|w_\varepsilon'(t)| \leq K\, \varepsilon^{-\frac{1}{2m}} \quad for\ m \geq 1 \ .$$

P3: For all $t \in [t_1, t_2]$ and $m \geq 1$ it is $t\, w_\varepsilon'(t) \geq 0$.

Setting $A = \inf\limits_{\Omega} \frac{b^*(x)}{a_{nn}(x)} > 0$ we can prove that

$$S_\varepsilon(x) \equiv \begin{cases} \frac{2}{\alpha_0}\, \varepsilon^{\frac{1}{m}-1} (K - w_\varepsilon(x_n))[f_\varepsilon]_0 & \text{for } m > 1 \\ \frac{2}{\alpha_0} (K\, |\ln \varepsilon| - w_\varepsilon(x_n))[f_\varepsilon]_0 & \text{for } m = 1 \end{cases}$$

is a barrier function for u_ε and consequently

$$[u_\varepsilon]_0 \leq K\, \varepsilon^{\frac{1}{m}-1} [f_\varepsilon]_0 \qquad \text{for } m > 1\,,$$
$$[u_\varepsilon]_0 \leq K\, |\ln \varepsilon|\, [f_\varepsilon]_0 \qquad \text{for } m = 1\,.$$

In the case $b^*(x) > 0$ in general it is impossible to obtain estimates of the same kind. For instance, for the solution of the boundary layer problem

$$-\varepsilon\, \Delta u_\varepsilon + x_1 \frac{\partial u_\varepsilon}{\partial x_1} + x_2 \frac{\partial u_\varepsilon}{\partial x_2} = 1 \qquad \text{in } |x| < 1,$$
$$u_\varepsilon = 0 \qquad \text{on } |x| = 1$$

there exist positive constants such that for sufficiently small ε

$$[u_\varepsilon]_0 \geq \gamma_1 \exp\left(\frac{\gamma_2}{\varepsilon}\right)\,.$$

Analogous behaviour of the solution u_ε is also possible in the case $m > 1$.

(iv) the case $b_n(x) = x_n^l\, b^*(x)$, $b^*(x) \neq 0$, l even, $l \geq 2$.
Without restriction of generality let $b^*(x) \leq -b_0 < 0$ and $l = 2m$.
The function

$$v_\varepsilon(t) = \int\limits_{-\infty}^{t\varepsilon^{-\frac{1}{2m+1}}} \int\limits_{-\infty}^{q} \exp\left(\frac{A}{2m+1}(p^{2m+1} - q^{2m+1})\right) dp\, dq$$

has the following properties for every $A > 0$ and sufficiently small ε

P4: $$\varepsilon\, v_\varepsilon''(t) + A t^{2m} v_\varepsilon'(t) = \varepsilon^{1-\frac{2}{2m+1}}\,.$$

P5: For all $t \in [t_1, t_2]$ it is

$$0 \leq v_\varepsilon(t) \leq K , \qquad 0 \leq v'_\varepsilon(t) \leq K\varepsilon^{-\frac{1}{2m+1}} .$$

Setting $A = \inf_{\Omega} \frac{b^*(x)}{a_{nn}(x)} > 0$ we can prove that

$$S_\varepsilon(x) \equiv \frac{2}{\alpha_0}\, \varepsilon^{\frac{2}{2m+1}-1}\left(K - v_\varepsilon(x_n)\right)[f_\varepsilon]_0$$

is a barrier function for u_ε and consequently

$$[u_\varepsilon]_0 \leq K\varepsilon^{\frac{2}{2m+1}-1}[f_\varepsilon]_0 , \qquad m \geq 1 .$$

(v) the case $b_n(x) \equiv 0$.

Let $\gamma_1 = \max(1, \sup_{\Omega} \frac{1+a_n(x)}{a_{nn}(x)})$ and $\gamma_2 > 0$ with $\gamma_2 \geq \exp(-\gamma_1 x_n)$ in $\bar{\Omega}$.

Then the function

$$S_\varepsilon(x) \equiv \frac{1}{\gamma_1 \varepsilon}\left(\gamma_2 - \exp(-\gamma_1 x_n)\right)[f_\varepsilon]_0$$

majorizes u_ε and consequently

$$[u_\varepsilon] \leq \frac{K}{\varepsilon}[f_\varepsilon]_0 \quad .$$

(vi) the case $b_n(x) \equiv 0$, $b_1(x) = x_n^l\, b^*(x)$, $b^*(x) \neq 0$, $\bar{\Omega} \cap \{x \mid x_n < 0\} = \phi$.
Without restriction of generality let $b^*(x) > 0$. $S_{p,q}(z)$ denotes the Lommel function (BATEMAN/ERDELYI (1973,74)) and $\varkappa = \frac{1}{2+l} \in (0,1)$.
The function

$$v_\varepsilon(t) = (1-\varkappa)^{-\frac{1+3\varkappa}{2}}\, \varepsilon^{-\frac{1+3\varkappa}{4}}\, t^{\frac{1}{2}}\, S_{\frac{1-3\varkappa}{2},\frac{1-\varkappa}{2}}\left((1-\varkappa)\, t^{\frac{1}{1-\varkappa}}\, \varepsilon^{-\frac{1}{2}}\right)$$

has the following properties

P6: $\quad \varepsilon v''_\varepsilon(t) + t^l v_\varepsilon(t) = 1$.

P7: For all $t \in [0, t_2]$ it is

$$|v_\varepsilon(t)| \leq K\varepsilon^{-\varkappa}, \qquad |v'_\varepsilon(t)| \leq K\varepsilon^{-\frac{1+\varkappa}{2}} .$$

For appropiately chosen positive constants K_1, K_2, K_3 we can prove that

$$S_\varepsilon(x_1, x_n) = K_1[f_\varepsilon]_0 \exp(K_2 x_1)\left(K_3\varepsilon^{-\varkappa} + v_\varepsilon(x_n)\right)$$

is a barrier function for u_ε and consequently

$$[u_\varepsilon] \leq K\,\varepsilon^{-\varkappa}[f_\varepsilon]_o \quad .$$

For inhomogeneous boundary value problems of the form

$$\begin{aligned} \varepsilon L_1 u_\varepsilon + L_o u_\varepsilon &= f_\varepsilon \qquad && in\ \Omega \\ u_\varepsilon &= g_\varepsilon && on\ \partial\Omega \end{aligned} \tag{2.12}$$

the functions $S_\varepsilon + \sup\limits_{x\in\partial\Omega} |\ g_\varepsilon(x)|$ majorize the solution u_ε of (2.12) in the cases (i) - (vi). We summarize the results in

Theorem 2.10. Let the assumptions of Theorem 2.3 be fulfilled. L_o satisfies the assumptions of one of the cases (i) - (vi). Then the a-priori estimate

$$[u_\varepsilon]_o \leq K\,\varepsilon^{-\varkappa}[f_\varepsilon]_o + \sup_{x\in\partial\Omega} |g_\varepsilon(x)|$$

is valid for any solution u_ε of (2.12) and the constant $\varkappa \in [0,1]$ can explicitly be given. ∎

Remark. If it is possible to transform the operator L_o into $\widetilde{L}_o$ by an one-to-one mapping $x \to (\varrho, \varphi_1, \ldots, \varphi_{n-1})$ regular on $\bar{\Omega}$ such that $b \cdot \nabla\varrho = \varrho^l b^*$, $b^* \neq 0$ and $b \cdot \nabla\varrho \equiv 0$, respectively, we can use $S_\varepsilon(\varrho(x))$ as barrier function and Theorem 2.10 remains true. This will be applied in section 4.5. ∎

2.4. A-priori estimates for singularly perturbed parabolic problems

As in the elliptic case we first consider problems of arbitrary order of the form

$$\begin{aligned} \frac{\partial u_\varepsilon}{\partial t} + \varepsilon L_1 u_\varepsilon + L_o u_\varepsilon &= f_\varepsilon \qquad && in\ Q \\ \frac{\partial^s u_\varepsilon}{\partial \nu^s} &= 0 && on\ S_T \\ u_\varepsilon &= 0 && in\ \Omega \end{aligned} \tag{2.13}$$

where $L_1 \equiv \sum\limits_{|\beta|\leq 2m} a_\beta(x)\, D^\beta$, $\quad L_o \equiv \sum\limits_{\mu=1}^{n} b_\mu(x)\, \dfrac{\partial}{\partial x_\mu} + c(x)$.

Let the coefficients of L_1 and L_0, the boundary $\partial\Omega$ of Ω and f_ε be sufficiently regular. Then it follows

<u>Theorem 2.11.</u> Let the assumptions of Theorem 2.2 be fulfilled for the problem (2.13). Then the solution u_ε of (2.13) satisfies the a-priori estimate

$$\|u_\varepsilon\| \leq K\varepsilon^{1+\frac{l^2}{(2m-1)(2m-1-\alpha)}-\frac{l+\alpha}{2m}} \|f_\varepsilon\|_{l-2m+\alpha,2m} +$$

$$+K\varepsilon^{-\frac{nl}{2(2m-1)(2m-1-\alpha)}-\frac{n}{2l}} \left(\int_0^T \int_\Omega f_\varepsilon^2 \, dx\, dt\right)^{\frac{1}{2}} . \quad \blacksquare$$

<u>Proof.</u> BESJES (1974) proved the a-priori estimate

$$\|u_\varepsilon\| \leq K\varepsilon^{1+\frac{l^2}{(2m-1)(2m-1-\alpha)}-\frac{l+\alpha}{2m}} \|f_\varepsilon\|_{l-2m+\alpha,2m} +$$

$$+K\varepsilon^{-\frac{nl}{(2m-1)(2m-1-\alpha)}-\frac{n}{2l}} \sup_t \left(\int_\Omega u_\varepsilon^2(x,t)\, dx\right)^{\frac{1}{2}}$$

for $l \geq 2m$ for any solution of (2.13). Without restriction of generality in the parabolic case there exists a K_0 with $c(x) - \frac{1}{2}\operatorname{div} b(x) \geq K_0 > 0$. So it follows from Garding's inequality that

$$\int_\Omega u(x,t)\frac{\partial u}{\partial t}dx + \int_\Omega u(x,t)(\varepsilon L_1 + L_0)u(x,t)dx \geq \frac{1}{2}\frac{\partial}{\partial t}\int_\Omega u_\varepsilon^2(x,t)\,dx + \frac{K_0}{2}\int_\Omega (u(x,t))^2 dx$$

and

$$\frac{1}{2}\frac{\partial}{\partial t}\left(\int_\Omega u_\varepsilon^2(x,t)\,dx\right) \leq K_1 \int_\Omega u_\varepsilon^2(x,t)\,dx + K_2 \int_\Omega f_\varepsilon^2(x,t)\,dx \quad .$$

We obtain

$$\int_\Omega u_\varepsilon^2(x,t)\,dx \leq K_3 \int_0^t \int_\Omega f^2(x,\tau)\exp(K_1(t-\tau))\,dx\,d\tau$$

and consequently, the assertion of the theorem . $\blacksquare$

For problems of second order the a-priori estimate can essentially be sharpened and the coefficients of L_1 and L_0 may also depend on t. Therefore we consider

$$\begin{aligned} \frac{\partial u_\varepsilon}{\partial t} + \varepsilon L_1 u_\varepsilon + L_0 u_\varepsilon &= f_\varepsilon && \text{in } Q \\ u_\varepsilon &= g_\varepsilon && \text{on } S_T \\ u_\varepsilon &= h_\varepsilon && \text{in } \Omega \end{aligned} \tag{2.14}$$

where $L_1 \equiv \sum_{|\beta|\le 2} a_\beta(x,t)\, D^\beta, \qquad L_0 \equiv \sum_{\mu=1}^{n} b_\mu(x,t) \frac{\partial}{\partial x_\mu} + c(x,t)$.

Without restriction of generality there is a positive constant K_0 with $c(x,t) \geq K_0$.

Theorem 2.12. Let the assumptions of Theorem 2.5 be fulfilled for (2.14). Then the a-priori estimate

$$\|u_\varepsilon\| \leq K\|f_\varepsilon\| + \sup_{S_T}|g_\varepsilon(x,t)| + \sup_{\Omega}|h_\varepsilon(x)|$$

is valid. ∎

Proof. For $K = \frac{2}{K_0}$ the constant on the right hand side of the estimate is a barrier function for u_ε. The maximum principle yields the assertion of Theorem 2.12. ∎

2.5. Application of Banach's fixed point theorem. Estimates of the solution of special elliptic problems of second order

To prove local existence and uniqueness of the solutions of special quasilinear elliptic problems of second order we use a modification of Banach's fixed point theorem. Let N be a nonlinear operator with $N(0) = 0$ mapping the Banach-space X into the Banach-space Y. For the decomposition

$$Nu = Lu + Ru$$

let the assumptions

(H7) $L: X \to Y$ linear and one-to-one, L^{-1} continuous with $\|u\|_X \leq \frac{1}{I}\|Lu\|_Y$

(H8) R is bounded Lipschitz-continuous, that means there is a $\sigma^* > 0$ such that for all $\sigma \in [0,\sigma^*]$ and all $u_1, u_2 \in B(0,\sigma) = \{u \in X \mid \|u\| \leq \sigma\}$

$$\|Ru_1 - Ru_2\|_Y \leq m(\sigma)\|u_1 - u_2\|_X$$

is valid where $m(\cdot)$ is monotonic nondecreasing and $\lim_{\sigma\to 0} m(\sigma) = 0$.

be fulfilled. If N is differentiable at the point 0 in the sense of Frechèt, then we have $L = N'(0)$.

Theorem 2.13. For every $f \in Y$ with $\|f\|_Y \leq \frac{1}{2}\sigma_0$ and $\sigma_0 = \sup\{\sigma \in [0,\sigma^*] | m(\sigma) \leq \frac{1}{2}\}$ there exists exactly one solution of the equation

$$Nu = f$$

in $B(0,\sigma_0)$ and the estimate

$$\|u\|_X \leq \frac{2}{l}\|f\|_Y$$

is valid. ∎

Proof. Because of the existence of L^{-1}, the equation $Nu = f$ is equivalent to

$$u = Tu \equiv L^{-1}(f - Ru) \; .$$

We shall now indicate that T fulfills the assumptions of Banach's fixed point theorem on every ball $B(0,\sigma)$ with $\frac{2}{l}\|f\|_Y \leq \sigma \leq \sigma_0$.

(i) $T: B(0,\sigma) \longrightarrow B(0,\sigma)$.
From (H7) and (H8) it follows first that

$$\|Tu\|_X \leq \frac{1}{l}\|f - Ru\|_Y \leq \frac{1}{l}\left(\|f\|_Y + m(\sigma)\cdot\sigma\right) .$$

According to the definition of σ_0, the monotonicity of $m(\cdot)$ and $\|f\|_Y \leq \frac{1}{2}\sigma$ we obtain

$$\|Tu\|_X \leq \frac{\sigma}{2} + \frac{\sigma}{2} \leq \sigma \quad .$$

(ii) T is contractive.
In analogy to (i) we obtain

$$\|Tu_1 - Tu_2\|_X \leq \frac{1}{l} m(\sigma)\|u_1 - u_2\|_X$$

and consequently the contractibility of T for $\sigma \leq \sigma_0$.

Setting $\sigma = \sigma_0$ we obtain the existence and uniqueness of a solution of $Nu = f$ in $B(0,\sigma_0)$. The choice $\sigma = \frac{2}{l}\|f\|_Y$ yields the a-priori estimate . ∎

In section 4.6 we shall apply Theorem 2.13 in the following way. For the problem $A_\varepsilon u_\varepsilon = 0$ let a function ϕ_ε be known with $A_\varepsilon \phi_\varepsilon = f_\varepsilon$ and $\lim\limits_{\varepsilon \to 0} \|f_\varepsilon\| = 0$. Then we set $z_\varepsilon = u_\varepsilon - \phi_\varepsilon$, $N z_\varepsilon \equiv A_\varepsilon(z_\varepsilon + \phi_\varepsilon) - A_\varepsilon \phi_\varepsilon$ and (H7), (H8) and $\|f_\varepsilon\|_Y \leq \frac{1}{2}\sigma_0$ need to be proved. If A_ε is differentiable at the point ϕ_ε in the sense of Frechêt, then we have $L = A'_\varepsilon(\phi_\varepsilon)$. Therefore, to prove (H7) we have to exend the Theorems 2.8 and 2.10 to linear elliptic problems with special dependences of the coefficients on ε .

We now consider

$$\tilde{L}_\varepsilon u \equiv \varepsilon L u + \sum_{\mu=1}^{n} b_\mu(x,\varepsilon) \frac{\partial u}{\partial x_\mu} + c(x,\varepsilon) u$$

under the conditions

(H9) L is a linear uniformly elliptic differential operator of second order with regular coefficients being independent of ε .

(H10) let b_μ, c be regular and let the estimates

$$\sum_{\mu=1}^{n} \left([b_\mu]_0 + \varepsilon^\alpha [b_\mu]_\alpha \right) \le K$$
$$[c]_0 + \varepsilon^\alpha [c]_\alpha \le K \varepsilon^{-1}$$

be fulfilled for sufficiently small ε.

<u>Theorem 2.14.</u> Assuming (H 9) and (H10) the a-priori estimate

$$|u|_{2+\alpha} \le K(\varepsilon^{1+\alpha} \| \tilde{L}_\varepsilon u \|_\alpha + [u]_0)$$

is valid for all $u \in C_0^{2+\alpha}(\bar{\Omega})$ with a constant K being independent of ε.■

<u>Proof.</u> Applying Theorem 2.1 to the boundary value problem

$$Lu = \frac{1}{\varepsilon}\left(\tilde{L}_\varepsilon u - \sum_{\mu=1}^{n} b_\mu \frac{\partial u}{\partial x_\mu} - c u \right) \qquad in\ \Omega,$$
$$u = 0 \qquad on\ \partial\Omega$$

we obtain

$$[u]_{2+\alpha} \le \|u\|_{2+\alpha} \le \frac{K}{\varepsilon} \| \tilde{L}_\varepsilon u \|_\alpha + \frac{K}{\varepsilon} \| \sum_\mu b_\mu \frac{\partial u}{\partial x_\mu} \|_\alpha + \frac{K}{\varepsilon} \| c u \|_\alpha + [u]_0 .$$

Because of (H10) the relations

$$\frac{K}{\varepsilon} \| \sum_\mu b_\mu \frac{\partial u}{\partial x_\mu} \|_\alpha \le K \varepsilon^{-1} \|u\|_{1+\alpha} + K \varepsilon^{-1-\alpha} [u]_1$$
$$\frac{K}{\varepsilon} \| c u \|_\alpha \le K \varepsilon^{-2} \|u\|_\alpha + K \varepsilon^{-2-\alpha} [u]_0$$

are satisfied and using Lemma 2.1 ($\omega = \omega_0$) we have

$$[u]_{2+\alpha} \le K\varepsilon^{-1} \| \tilde{L}_\varepsilon u \|_\alpha + K\varepsilon^{-1}[u]_{1+\alpha} + K\varepsilon^{-1-\alpha}[u]_1 + K\varepsilon^{-2}[u]_\alpha + K\varepsilon^{-2-\alpha}[u]_0$$

for sufficiently small ε .

Repeated application of Lemma 2.1 with $\omega = \frac{\varepsilon^{2+\alpha-p}}{4K}$ yields

$$K\varepsilon^{p-2-\alpha}[u]_p \le \frac{1}{4}[u]_{2+\alpha} + K\varepsilon^{-2-\alpha}[u]_0 , \quad 0 < p < 2+\alpha .$$

Summarizing, it follows

$$\varepsilon^{2+\alpha}\,[u]_{2+\alpha} \leq K\left(\varepsilon^{1+\alpha}\,\|\tilde{L}_\varepsilon u\|_\alpha + [u]_0\right)$$

and consequently, the assertion . ■

As in the linear case the estimate of $[u]_0$ depends on the structure of $b_\mu(x,\varepsilon)$ and $c(x,\varepsilon)$. This will be discussed in dependence on special ϕ_ε in section 4.6.

Comments to Chapter 2

ECKHAUS/DE JAGER (1966) were the first who applied the maximum principle to singularly perturbed elliptic problems. The technique of estimates for elliptic and parabolic problems of arbitrary order was introduced by BESJES (1974, 1975a,b). The problems treated there degenerate, in general, to elliptic and parabolic problems of smaller order, respectively. The application and extension of the method to reduced problems of first order was realized by ROOS (1978) and LUBE/TOBISKA (1980). TOBISKA (1981) constructed barrier functions for problems of second order which are adapted to the special structure of L_0 . In the work of VAN HARTEN (1978) we have first found the use of Banach's fixed point theorem and the introduction of ε -dependent norms in spaces of Hölder-continuous functions for the investigation of nonlinear elliptic problem. The generalized maximum principle was first applied by FELGENHAUER (1980) on a special singularly perturbed parabolic problem.

3. PROPERTIES OF SOLUTIONS OF GLOBAL PROBLEM FOR ELLIPTIC AND PARABOLIC BOUNDARY VALUE PROBLEMS

In Chapter 4 we shall construct asymptotic solutions of the elliptic boundary value problem

$$L_\varepsilon u_\varepsilon \equiv \varepsilon L_1 u_\varepsilon + L_0 u_\varepsilon = f_\varepsilon \qquad \text{in } \Omega$$
$$\frac{\partial^s u_\varepsilon}{\partial \nu^s} = 0 \qquad \text{on } \partial\Omega\,, \quad s = 0,1,\dots,m-1$$

and of the parabolic problem

$$\frac{\partial u_\varepsilon}{\partial t} + L_\varepsilon u_\varepsilon = f \qquad \text{in } Q$$
$$\frac{\partial^s u_\varepsilon}{\partial \nu^s} = 0 \qquad \text{on } S_T\,, \quad s = 0,1,\dots,m-1$$
$$u_\varepsilon = 0 \qquad \text{in } \Omega \quad .$$

The results of Chapter 1 show that the solutions of

$$L_0 u = f$$

and

$$\frac{\partial u}{\partial t} + L_0 u = f$$

with certain additional condition are of great importance.
Therefore we shall give now statements on existence and regularity of the solutions of boundary value problems for equations of first order.

We first consider

$$L_0 u \equiv \sum_{i=1}^{n} b_i(x) \frac{\partial u}{\partial x_i} + c(x) u = f \quad \text{in } \Omega \qquad\qquad (3.1)$$
$$u|_\Sigma = g$$

where Σ is a subset of the boundary $\Gamma = \partial\Omega$. In principle we suppose that the coefficients are regular in a domain which properly includes Ω and that f, g and Γ are also sufficiently regular. By solving (3.1) by the method of characteristics, the solutions of the dynamical system

$$\dot{\xi}(\tau) = b(\xi(\tau)) \qquad\qquad (3.2)$$

corresponding to (3.1) are essential, namely, it then follows that

$$\frac{du}{d\tau} = f(\xi(\tau)) - c(\xi(\tau))u \quad .$$

With the initial condition $\xi(0) = x \in \overline{\Omega}$ (3.2) is locally uniquely solvable and the solution $\xi_x(\tau)$ (the characteristic through x) can be uniquely extended up to the boundary Γ and unbounded for $\tau \to \pm\infty$

in Ω , respectively.

Obviously, to determine Σ the behaviour of characteristics in the neighbourhood of Γ is essential. Let F(x) be a sufficiently regular function with $F|_{\Gamma} = 0$, $F < 0$ in Ω and with an uniformly non-vanishing gradient in the neighbourhood of the boundary (there is a neighbourhood Ω^* of Γ such that $\min\limits_{x\in\Omega^*} \sum\limits_{i=1}^{n} \left(\frac{\partial F}{\partial x_i}\right)^2 \geq \gamma > 0$). Then the characteristics intersect the boundary with non-vanishing angle at $\Gamma_{\pm}$ with

$$\Gamma_{\pm} = \{ x \in \Gamma \mid b\cdot \nabla F \gtrless 0\}$$

and are tangent to it on Γ_0 with

$$\Gamma_0 = \{ x \in \Gamma \mid b\cdot \nabla F = 0 \} \qquad \text{(see fig. 3.1).}$$

In the case of a regular boundary it holds that $\Gamma = \Gamma_0 \cup \Gamma_+ \cup \Gamma_-$.

It is obvious to impose additional conditions on $\Sigma = \overline{\Gamma}_-$ or $\Sigma = \overline{\Gamma}_+$. In the construction of asymptotic solutions in Chapter 4 local correctors are determined in the neighbourhood of $\Gamma \setminus \Sigma$. The corresponding local problems must possess solutions of boundary layer type and it appears that this is not the case in the neighbourhood of Γ_- (see Appendix). Therefore we have to impose additional conditions on $\overline{\Gamma}_-$, consequently, we consider

$$\begin{aligned} L_0 u &= f \quad \text{in } \Omega \\ u|_{\overline{\Gamma}_-} &= g \quad . \end{aligned} \tag{3.1$_*$}$$

First we have

<u>Lemma 3.1.</u> If $c(x) > 0$ in $\overline{\Omega}$, then the solution of $(3.1)_*$ is unique. ∎

<u>Proof.</u> Let $K = \max\{\sup\limits_{\Omega} |\frac{f}{c}| + \max\limits_{\overline{\Gamma}_-} |g|\}$ and $v_{\pm} = K \pm u$.

Then it follows that

$$L_0(v_{\pm}) = K\cdot c \pm f \leq 0 \qquad \text{in } \overline{\Omega} \quad .$$

We suppose $v_{\pm}$ has a negative minimum at $x_0 \in \Omega$. Then from $\left.\frac{\partial v_{\pm}}{\partial x_i}\right|_{x_0} = 0$ and $v_{\pm}(x_0) < 0$ it follows the contradiction $L_0(v_{\pm}) > 0$ at x_0. Analogously, it can be shown that $v_{\pm}$ cannot have a negative minimum at $x_0 \in \Gamma \setminus \overline{\Gamma}_-$. From $v_{\pm} \geq 0$ on $\overline{\Gamma}_-$ we finally obtain $v_{\pm} \geq 0$ in $\overline{\Omega}$ and

$$\max_{\overline{\Omega}} |u| \leq \max\{ \sup_{\Omega} |\frac{f}{c}| , \max_{\overline{\Gamma}_-} |g|\} \; .$$

The uniqueness follows immediately. ■

In the case of $c(x) \equiv 0$ the solution of $(3.1)_*$ is not unique in general, for instance, if $\bar{\Gamma}_-$ is the empty set we can conclude that any constant is a solution.

From simple examples we can already see that the regularity of the solution depends essentially on $c(x)$. Considering

$$-x_1 \frac{\partial u}{\partial x_1} - x_2 \frac{\partial u}{\partial x_2} + c\,u = 1 \quad in\ \Omega = \{(x_1, x_2) \mid |x| = (x_1^2 + x_2^2)^{\frac{1}{2}} < 1\}$$

$$u|_\Gamma = 0 \ ,$$

we have $u = \frac{1}{c}(1 - |x|^c)$.

On the other hand, discontinuities appear if, for instance, there are points $y \in \bar{\Gamma}_- \cap \Gamma_0$ such that $\xi_y(\tau)$ is contained in the interior of Ω. In this manner the solution of

$$\frac{\partial u}{\partial x_2} + \lambda u = 0 \qquad in\ \ \Omega = \{(x_1, x_2) \mid 1 < |x| < 2\}$$

$$u\Big|_{\substack{|x|=2\\ x_2 \le 2}} = 0$$

$$u\Big|_{\substack{|x|=1\\ x_2 \ge 0}} = 1$$

becomes discontinuous along $x_1 = \pm 1$, $0 \le x_2 \le \sqrt{3}$.

The simplest case appears if Γ_- is closed in Γ (then we have $\bar{\Gamma}_- \cap \Gamma_0 = \emptyset$). For a simply connected domain Ω we then obtain $\Gamma = \Gamma_-$ or $\Gamma_- = \emptyset$ while in the case of a multiple connected domain this statement is true for the connected components of Γ. Under the condition that Γ_- is closed in Γ, we have the surprising result that for sufficiently large $c(x)$ the solution of $(3.1)_*$ is regular independent of the behaviour of characteristics, more precisely, it holds

Theorem 3.1. (OLEINIK/RADKEWICZ (1971))

Let Γ_- be closed in Γ. Then for $c(x) > c_0(n, \kappa, \Omega)$ the solution u of $(3.1)_*$ belongs to the space $C^{\kappa,\alpha}(\bar{\Omega})$ with arbitrary α, $0 \le \alpha < 1$, where

$$c_0 = max\{0, \max_{j=0,\dots,\kappa+1} \beta_j\}$$

$$\beta_0 = \sup_{x \in \Omega}[-div\ b(x)]$$

$$\beta_l = \sup_{x \in \Omega}\Big[\max_{T_l}\Big\{\sum_{j=1}^{l} \frac{\partial b_{s_j}}{\partial x_{s_j}} + \frac{1}{2}\sum_{j=1}^{l}\sum_{\substack{\mu=1\\ (\mu \ne s_j)}}^{n}\Big(\Big|\frac{\partial b_\mu}{\partial x_{s_j}}\Big| + \Big|\frac{\partial b_{s_j}}{\partial x_\mu}\Big|\Big\}\Big]$$

$$T_l = (s_1, \dots, s_l)\ , \qquad s_j = 1, \dots, n\ , \qquad l = 1, \dots, \kappa+1 \quad .$$ ■

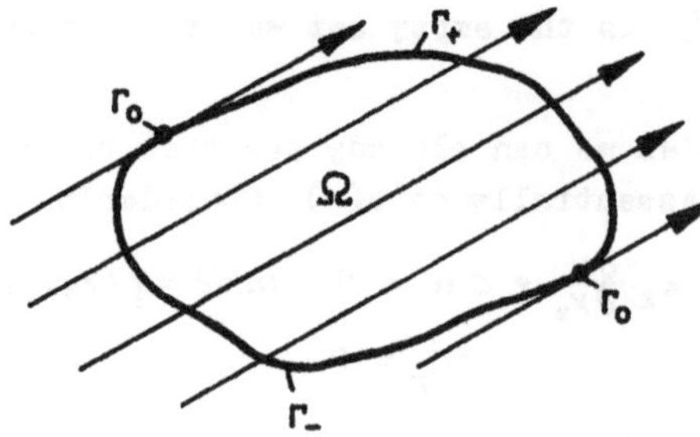

Fig. 3.1

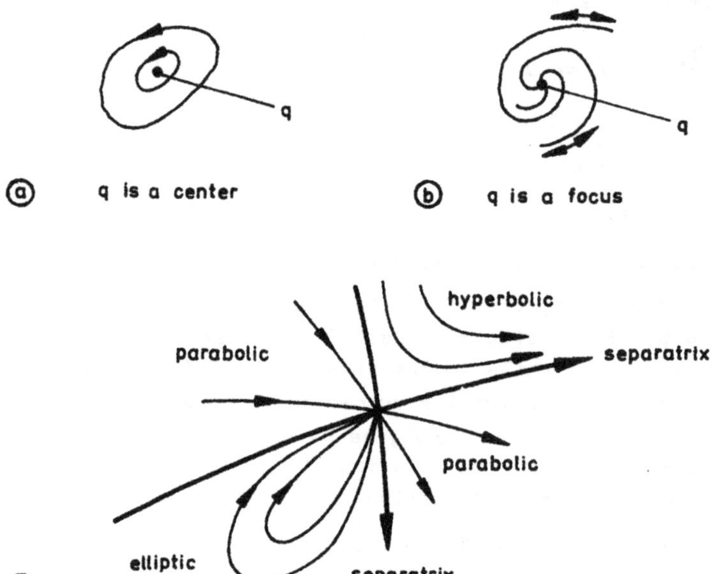

Fig. 3.2

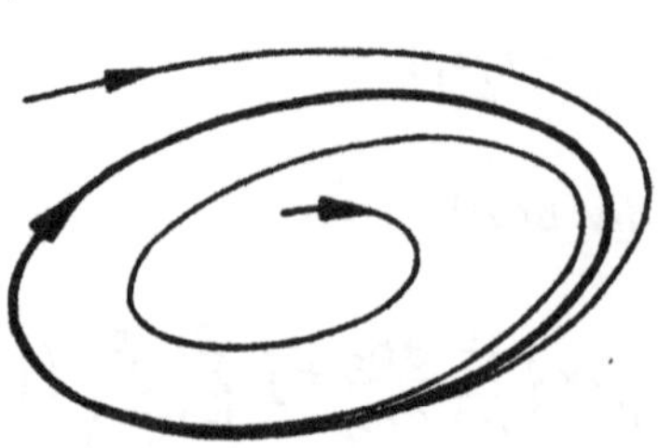

Fig. 3.3

The assumptions with respect to the absolute term can be weakened in the case of a simple behaviour of characteristics. First of all, let for $x \in \Omega$

$$\omega(x) = \bigcap_{0 \le s < \infty} \{ \xi_x(\tau) \mid s \le \tau < \infty \}$$

$$\alpha(x) = \bigcap_{-\infty < s \le 0} \{ \xi_x(\tau) \mid -\infty < \tau \le s \} \quad .$$

It is usual to say that the characteristic $\xi_x(\tau)$ through x originates at $\alpha(x)$ and dies at $\omega(x)$. Let $I(\bar{\Omega})$ denote the union of all the limit sets of the above type entirely contained in $\bar{\Omega}$. The stationary points x^* of the field b(x), that means the points x^* with $b(x^*) = 0$, belong to $I(\bar{\Omega})$. If Ω is simply connected and Γ_- is closed in Γ, then there is at least one stationary point.

The local structure of characteristics in the neighbourhood of stationary points varies for $n \ge 2$. In the case n = 2, for instance, we can topologically classify an isolated stationary point x^* as follows:

(i) all characteristics of (3.2) are closed, x^* is a center (see fig. 3.2a)

(ii) all characteristics of (3.2) spiral towards to (or away from) x^*, x^* is a focus (see fig. 3.2b)

(iii) near x^* there is a finite number of elliptic, hyperbolic and parabolic sectors (see fig. 3.2c).

In the case of κ-dimensional ($\kappa = 1,\ldots,n$) connected stationary manifolds in $\bar{\Omega}$ an appropriate classification is a difficult problem.

Closed characteristics, that means the images of periodic solutions of (3.2) are another example for subsets of $I(\bar{\Omega})$. They can appear, for instance, as continuum (see fig. 3.2a) or as an isolated limit-cycle (see fig. 3.3).

$I(\bar{\Omega})$ is empty if and only if all characteristics leave $\bar{\Omega}$ in finite time. It is interesting that the following lemma holds.

Lemma 3.2. (DEVINATZ/FRIEDMAN (1974))
All characteristics of $\bar{\Omega}$ leave $\bar{\Omega}$ in finite time if and only if there exists a regular function $\eta(x)$ with

$$b \cdot \nabla \eta \ge b_0 > 0 \quad in \ \bar{\Omega} \ . \qquad \blacksquare$$

From this fact it follows

Theorem 3.2. Let Γ_- be closed in Γ and let all characteristics leave $\bar{\Omega}$ in finite time. Then the solution u of $(3.1)_*$ belongs to the space $C^\kappa(\bar{\Omega})$ for any c(x). $\blacksquare$

The proof simply follows from

$$L_0 v = e^{\gamma\eta}\left[b\cdot\nabla v + (c(x) + \gamma\, b\cdot\nabla\eta)v\right]$$

satisfied for the function $v = \exp(-\gamma\eta(x))u$. Choosing γ appropriately, Theorem 3.1 can be applied.

From the above remarks on the existence of stationary points it is clear that Theorem 3.2 is meaningless for simple connected domains.

It was already shown by an example that if there are stationary points, then the regularity of solution depends on c(x). The following example shows that the regularity of the solution cannot be guaranteed for arbitrary c even in the case when Γ_- is closed in Γ and there are no stationary points.

We consider

$$[x_1(x_1^2+x_2^2-1) - x_2(x_1^2+x_2^2+1)]\frac{\partial u}{\partial x_1} + [x_2(x_1^2+x_2^2-1) + x_1(x_1^2+x_2^2+1)]\frac{\partial u}{\partial x_2} + cu = f(x)$$

in the domain

$$\left(\frac{1}{2}\right)^2 < x_1^2 + x_2^2 < 2^2 .$$

Choosing

$$F = \frac{1}{2}(x_1^2 + x_2^2) - K_{1,2}$$

we obtain

$$b\cdot\nabla F = (x_1^2 + x_2^2)(x_1^2 + x_2^2 - 1)$$

and moreover it is

$$b_1^2 + b_2^2 = 2(x_1^2 + x_2^2)\left(1 + (x_1^2 + x_2^2)^2\right) .$$

Consequently, Γ_- is closed in Γ and there are no stationary points. However, $x_1^2 + x_2^2 = 1$ is a closed characteristic and $u(\xi(\tau))$ is periodic on the unit circle. Setting $u(\xi(\tau)) = u(\tau)$, $c(\xi(\tau)) = c(\tau)$ and $f(\xi(\tau)) = f(\tau)$ from

$$\frac{du}{d\tau} + c(\tau)u(\tau) = f(\tau)$$

and from existence of a periodic solution with $u(0) = u(\tau_0)$ we obtain

$$u(0)\left(1 - \exp\int_0^{\tau_0} c(\tau)\,d\tau\right) = -\int_0^{\tau_0} F(\tau)\left[\exp\int_0^{\tau} c(\gamma)\,d\gamma\right]d\tau .$$

From this it follows that there exists a solution only in the case $\int_0^{\tau_0} c(\tau)\, d\tau \neq 0$. Then u(0) is not determined by the boundary conditions but only by b, c and f.

Remark. If c(x) is positive, hence $c(x) \geq c_0 > 0$, then u is almost everywhere continuous according to STROOCK/VARADHAN (1971). When, additionally, $(\bar{\Gamma}_- \cap \bar{\Gamma}_+) \cap \Gamma_0 = \emptyset$ we obtain the continuity of u. ∎

In case Γ_- is not closed in Γ, statements on regularity, being globally valid in $\bar{\Omega}$ cannot be expected. Therefore in the following we shall study regularity in subsets. For a point $x \in \Omega \cup \bar{\Gamma}_-$ let

$$\tau_+(x) = \inf\{\tau > 0 \mid \xi_x(\tau) \notin \Omega\}$$

be the exit time from Ω. For $\Gamma' \subset \bar{\Gamma}_-$ we call

$$G(\Gamma') = \{\xi_x(\tau) \mid x \in \Gamma',\ 0 \leq \tau \leq \tau_+(x)\}$$

the domain of influence of Γ'.

Theorem 3.3. Let Σ be a connected compact set with $\bar{\Sigma} \subset \Gamma_-$. Then the problem

$$L_0 u = f \qquad in\ G(\Sigma)$$
$$u = g \qquad on\ \bar{\Sigma}$$

has an unique and sufficiently regular solution u in $G(\Sigma) \setminus U(I(\bar{\Omega}))$ where U is an open neighbourhood of $I(\bar{\Omega})$. ∎

Proof. According to the definition of the domain of influence u is unique in $G(\Sigma) \setminus U(I(\bar{\Omega}))$. From the theory of ordinary differential equations, first of all, we have only a local result on existence and regularity of the solution of the system equivalent to $(3.1)_*$

$$\dot{x} = b$$
$$\dot{u} = f - cu$$

and consequently also for u. To prove the asserted global result we use appropriate compactness arguments.

First we shall show that there exists a sufficiently regular solution u in some neighbourhood of the image $\mathcal{E}_x = \{\xi_x(\tau) \mid x \in \Sigma\ ,\ 0 \leq \tau \leq \tau_0\}$ of each characteristic ξ_x with $\tau_0 \leq \tau_+(x) < +\infty$ and arbitrary $\tau_0 < +\infty$ for $\tau_+(x) = +\infty$, respectively.

Because $\mathcal{E}_x$ does not contain stationary points, the field b(x) is diffeomorphic to the constant field $e_1(x) = (1,0,\ldots,0)$ in a sufficiently small neighbourhood $\mathcal{V}(x_0)$ of an arbitrary point $x_0 \in \mathcal{E}_x$, that means there is a sufficiently regular bijection

$$\mathcal{F} : \mathcal{V}(x_0) \longrightarrow \mathcal{F}(\mathcal{V}(x_0))$$

such that b(x) is locally straightened (see ARNOLD (1979), Theorem 2.1.1.). Since $\mathcal{E}_x$ is compact, there exists a finite open covering of $\mathcal{E}_x$ by balls K_i with $K_i \subset G(\Sigma)$ and centres x_i on $\xi_x(\tau)$ such that in each K_i the field b(x) can be locally straightened by a diffeomorphism $\mathcal{F}_i$.
Without restriction of generality let $x_i = \xi_x(\tau_i)$ and $0 = \tau_1 < \tau_2 < \dots$. Consequently, $K_i \cap K_{i+1} \neq \emptyset$ (see fig. 3.4).

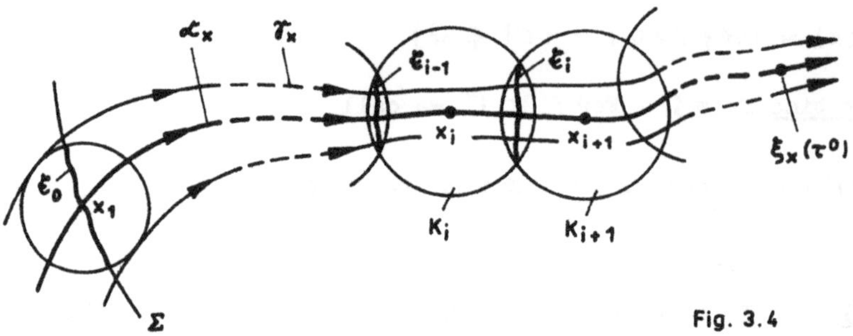

Fig. 3.4

According to a result of LEFSCHETZ (1957) (Chapter II, (17.4)) $\mathcal{E}_x$ can be imbedded in a characteristic-cylinder φ_x, for instance, we choose $\varphi_x = \{ \xi_{x'}(\tau) \mid x' \in K_1 \cap \Sigma \,,\ 0 \le \tau \le \tau^0 \}$ with sufficiently large τ^0 and sufficiently small ball K_1 around $x_1 = x \in \Sigma$. Then it follows $\varphi_x \subseteq \bigcup_i K_i$. Furthermore, there are sufficiently regular $(n-1)$-dimensional manifolds $\mathcal{E}_i$ with $\mathcal{E}_i \subseteq K_i \cap K_{i+1}$ and $\mathcal{E}_i$ transversal to the characteristics $\xi_{x'}(\tau)$ in φ_x (see LEFSCHETZ (1957), Chapter II, (16.5)). Further we suppose $\mathcal{E}_0 \subset \bar{\Sigma}$. From a theorem of KAMKE (1965) (D §1,4.3) we obtain the existence and regularity of a solution on $\varphi_x \cap K_1$ of the problem transformed by the diffeomorphism $\mathcal{F}_i$. Thereby the essential assumption is that one coefficient of the transformed field does not vanish in the considered domain. Consequently, we have a regular solution u_1 on K_1. The constructed solution u_1 implies a new boundary condition on $\mathcal{E}_1 \subseteq K_1 \cap K_2$. Analogously we obtain sufficiently regular solutions u_i on each ball K_i by induction. We note that because of the uniqueness $u_i \equiv u_{i+1}$ on $K_i \cap K_{i+1}$. Applying the partition of unit we conclude the existence of a sufficiently regular solution u in φ_x.

It is clear that $G(\bar{\Sigma}) \setminus U(I(\bar{\Omega}))$ is compact. So, from the covering by open characteristic cylinders we can choose a finite number of open characteristic cylinders $\varphi^{(i)}$ which are sufficient to cover

$$G(\bar{\Sigma}) \setminus U(I(\bar{\Omega})) \,.$$

The partition of the unit finally yields a regular solution on the domain under consideration . ∎

The statement of Theorem 3.3 is not true on the set $I(\bar{\Omega})$, even if it belongs to the domain of influence of Γ_-. In the case of

$$-x_1 \frac{\partial u}{\partial x_1} - x_2 \frac{\partial u}{\partial x_2} = 0 \qquad in \quad \Omega = \{(x_1, x_2) \mid |x| < 1\}$$
$$u = g \not\equiv const \quad on \quad \partial\Omega$$

we have $I(\bar{\Omega}) = \{0\}$, $\Gamma_- = \partial\Omega$ and $G(\Gamma_-) = \bar{\Omega}$ but the solution u is discontinuous at the point $x = 0$.

In the case of more complicated behaviour of the characterics we can apply the Theorems 3.1 and 3.3 in several subdomains. Let be given, for instance

$$(x_1^2 - 1)\frac{\partial u}{\partial x_1} + (x_2^2 - 1)\frac{\partial u}{\partial x_2} + cu = f \quad in \quad x_1^2 + x_2^2 < 3 .$$

Then it follows that

$$\bar{\Gamma}_- \cap \Gamma_0 = \{(-\sqrt{3}, \sqrt{3}), (\sqrt{3}, -\sqrt{3})\}$$

and

$$\Gamma_- = \{(x_1, x_2) \mid x_1^2 + x_2^2 = 3 , \; x_1 + x_2 < 0\} .$$

There are four stationary points (1,1), (1,-1), (-1,1), (-1,-1) in the considered domain.

Choosing

$$\Sigma = \{(x_1, x_2) \mid x_1^2 + x_2^2 = 3 , \; x_1 + x_2 < \gamma_1 < 0\} ,$$

for the domain of influence of Σ for sufficiently small γ_1 we obtain the domain

$$\Omega_1^* = \{(x_1, x_2) \mid x_1^2 + x_2^2 < 3 , \; x_1 < -1 - \gamma_2 , \; x_2 < -1 - \gamma_2 , \; x_2 < x_1 + 2\sqrt{3} - \gamma_2 ,$$
$$x_2 > x_1 - 2\sqrt{3} + \gamma_2\} , \; (\gamma_2 > 0 \;\; arbitrary)$$

On the other hand, let

$$\tilde{\Omega}_2^* = \{(x_1, x_2) \mid x_1^2 + x_2^2 < 3 , \; x_1 > -1 + \gamma_2 , \; x_2 > -1 + \gamma_2\}$$

and let Ω_2^* arise from $\tilde{\Omega}_2^*$ by rounding-off the corners. Then we can apply Theorem 3.1 in Ω_2^* for $c \geq c_0$. Consequently in the case of $c \geq c_0$ with sufficiently large c_0 we can guarantee the existence of a regular solution in $\Omega_1^* \cup \Omega_2^*$ for given boundary conditions on $\bar{\Sigma} \subset \Gamma_-$ (see fig. 3.5).

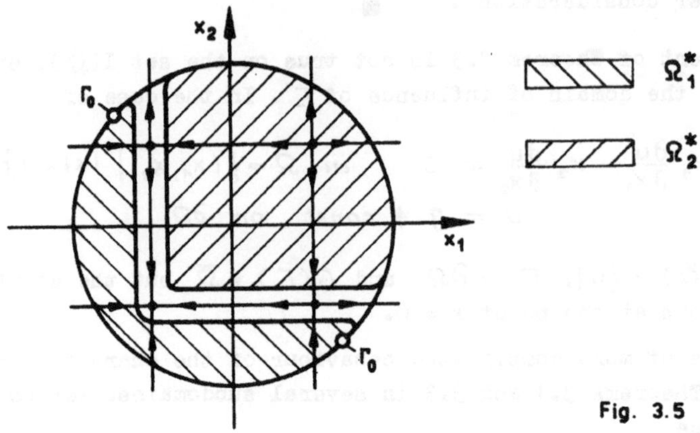

Fig. 3.5

Considering singularly perturbed parabolic problems we have to investigate

$$\frac{\partial u}{\partial t} + L_0 u = f \qquad in\ Q = \Omega \times (0,T)$$
$$u|_{\Sigma} = g \qquad (3.3)$$

where Σ is a subset of $\Omega \cup S_T$ with $S_T = \partial\Omega \times [0,T]$.

Setting $t = x_{n+1}$ and $x = (x_i)_{i=1}^{n+1}$, $b := (b,1)$ it seems that (3.3) is a simple special case of (3.1) since the characteristics through each point of Q can be extended up to the boundary, and the manifold $I(\bar{Q})$ defined as above is empty. As above we also define S_T^0, S_T^+, S_T^- and impose additional conditions on $\overline{\Omega \cup S_T^-}$.

So we consider

$$\frac{\partial u}{\partial t} + L_0 u = f \quad in \quad Q$$
$$u|_{\overline{\Omega \cup S_T^-}} = g \ . \qquad (3.3)_*$$

New difficulties arise in particular, when $\overline{\Omega \cup S_T^-}$ is not regular. Note that in difference to the above the domain Q is non-smooth.

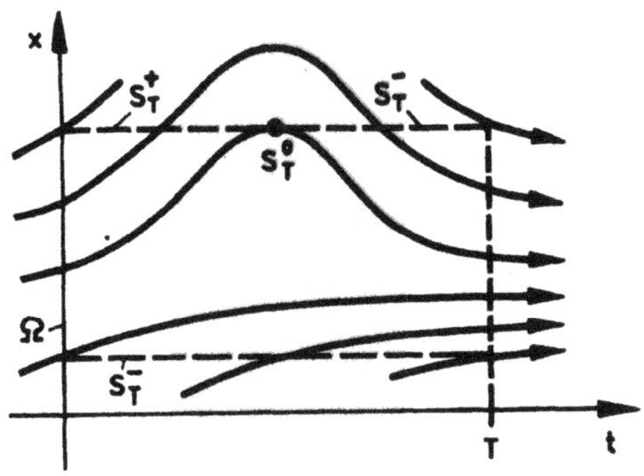

Fig. 3.6

The simplest case appears if S_T^- is empty since we only have to impose additional conditions on $\overline{\Omega}$.

Theorem 3.4. Let $S_T^- = \phi$. Then there exists a unique regular solution of

$$\frac{\partial u}{\partial t} + L_o u = f \qquad in \quad Q$$

$$u|_{\overline{\Omega}} = g \quad . \qquad \blacksquare$$

Proof. In the case $\Omega = \mathbb{R}^n$ the existence of a unique classical solution is a well-known result (see KAMKE (1965)).

Extending the coefficients and data appropriately, first of all we have a unique classical solution in $[0,T] \times \mathbb{R}^n$. We obtain the asserted result by restricting the solution to $\overline{Q}$, since because of $b \cdot \nu \geq 0$ on S_T every characteristic through a point of $\overline{Q}$ has passed a point $(0,x)$ with $x \in \overline{\Omega}$ only finite time τ before . $\blacksquare$

We now consider

$$\frac{\partial u}{\partial t} + \frac{\partial u}{\partial x} = f(x,t) \qquad in \ (0,1) \times (0,1)$$

$$u|_{t=0} = u|_{x=0} = 0 \quad .$$

Obviously, in this example S_T^- is closed in S_T and not empty, since $S_T^- = S_T \cap \{(x,t) \mid x = 0\}$.
It follows

$$u(x,t) = \begin{cases} \int_0^t f(\sigma + x - t, \sigma)\, d\sigma , & t \leq x \\ \int_{t-x}^t f(\sigma + x - t, \sigma)\, d\sigma , & x \leq t \end{cases}$$

consequently, u(x,t) is continuous, however, in general not continuously differentiable, for instance, we have

$$\frac{\partial u}{\partial x} = \begin{cases} \int_0^t f_x(\sigma+x-t,\sigma)\,d\sigma\,, & x > t \\ \int_{t-x}^t f_x(\sigma+x-t,\sigma)\,d\sigma + f(0,t-x)\,, & x < t \end{cases} .$$

u is continuously differentiable if and only if f(0,0) = 0.

Theorem 3.5.* Let S_T^- be closed in S_T and let g be continuous. Then the the solution u of $(3.3)_*$ belongs to

$$C(\overline{Q}) \cap H^1(Q) .$$

Only by satisfying certain compatibility conditions for f and g, the solution u becomes arbitrarily regular. ■

Proof. Defining, analogously as above, the domain of influence of an arbitrary subset of $\overline{\Omega \cup S_T^-}$, we obtain that as in Theorem 3.3 u is regular in the domains of influence $G(\overline{\Omega})$ of $\overline{\Omega}$ and $G(\overline{S_T^-})$ of $\overline{S_T^-}$. Every characteristic through a point of $\overline{S_T^-}$ has passed a point (0,x) only finite time before. The extention of $g|_{\overline{\Omega}}$ on the $\mathbb{R}^n$ defined in this way is continuous because of the continuity of g. Hence u is continuous.
From $u \in C(\overline{Q})$, $u \in C^1(Q_1)$, $u \in C^1(Q_2)$ with $\overline{Q_1 \cup Q_2} = \overline{Q}$ it follows that $u \in H^1(Q)$. The above example shows that in general u is not continuously differentiable. ■

If S_T^- is not closed in S_T we obtain statements on regularity in subsets by applying the Theorems 3.3 and 3.5.

Comments to Chapter 3

In the literature statements on existence and regularity of solutions of boundary value problems and initial-boundary value problems, respectively for equations of first order can be found, as a rule, as special cases of corresponding statements for equations of second order with nonnegative characteristic form. The central Theorem 3.1 is due to the basic work of OLEINIK/RADKEVIČ (1971) in which these problems are discussed.

A survey on the classical solvability of $(3.1)_*$ can be found in a work of YAMADA (1979), out of which we used the simple Lemma 3.1 and some examples.

Sufficient conditions for the existence of unique solutions $u \in L^2$ and, furthermore, conditions under which u has bounded variation can be found in BARDOS (1970, 1973). FRIEDRICHS (1958) considers systems of equations of first order.

*See also Lemma 4.1 p. 75

4. ASYMPTOTIC SOLUTIONS OF ELLIPTIC AND PARABOLIC BOUNDARY VALUE PROBLEMS

4.1. Global asymptotic solutions for equations of second order

We first consider the elliptic boundary value problem

$$L_\varepsilon u_\varepsilon \equiv -\varepsilon L_1 u_\varepsilon + L_0 u_\varepsilon = f \quad \text{in } \Omega$$
$$u_\varepsilon = 0 \quad \text{on } \partial\Omega \tag{4.1$_e$}$$

with

$$L_1 u \equiv \sum_{i,j} a_{ij} \frac{\partial^2 u}{\partial x_i \partial x_j} + \sum_i a_i \frac{\partial u}{\partial x_i} + a u$$
$$L_0 u \equiv \sum_1^n b_i \frac{\partial u}{\partial x_i} + c(x) u \quad .$$

In Chapter 3 we investigated the problem under which conditions the formal global solution, that means the solution of

$$L_0 u = f \ \text{ in } \Omega \ , \qquad u|_{\Gamma_-} = 0 \tag{4.2$_e$}$$

is regular.

In general, we now assume that, according to the Theorems 3.1, 3.2 and 3.3, at least it holds that $u \in C^2(\bar{\Omega})$ and $u \in C^2(G(\bar{\Sigma}))$, respectively and we study the question whether u is an uniformly asymptotic solution. Since u, in general, does not satisfy all the boundary conditions of $(4.1)_e$, u can be an uniformly asymptotic solution only in subdomains of Ω. Consequently, the subdomains $\Omega^* \subset \Omega$ are to be found such that

$$\lim_{\varepsilon\to 0} \sup_{x\in\Omega^*} | u_\varepsilon(x) - u(x) | = 0 \quad .$$

As in Chapter 3 we first assume that Γ_- is closed in Γ. Then we can expect that u is an uniformly asymptotic solution except for neighbourhoods of Γ_+ and Γ_0. Let $U_\gamma(\Gamma')$ (with $\Gamma' \subset \Gamma$) be the subset of Ω for which

$$dist(x, \Gamma') < \gamma \quad , \qquad (\gamma > 0) \quad .$$

Then it holds

Theorem 4.1. Let Γ_0 be empty and $c(x) \geq c_0 > 0$ such that $u \in C^2(\bar{\Omega})$. Then there exists a constant K independent of x, ε such that, for sufficiently small ε,

$$|u_\varepsilon(x) - u(x)| \leq K\varepsilon \quad \text{in} \quad \bar{\Omega} \setminus U_{\gamma(\varepsilon)}(\Gamma_+)$$

where $\gamma(\varepsilon)$ satisfies $\lim_{\varepsilon\to 0} \frac{\varepsilon \ln \varepsilon}{\gamma(\varepsilon)} = 0$. ■

<u>Proof.</u> Let $z_\varepsilon = u_\varepsilon - u$. Then it holds

$$|L_\varepsilon z_\varepsilon| \leq K_1 \varepsilon\,, \quad z_\varepsilon\big|_{\Gamma_-} = 0\,, \qquad |z_\varepsilon|_{\Gamma_+}| \leq K_2 \,.$$

According to Theorem 2.4 we now construct a barrier function majorizing z_ε. If Γ_+ is empty, then already $S = K_3\,\varepsilon$ is sufficient.
Let $F_+(x)$ be a regular function with the properties

$$F_+(x) = -\,dist(x, \Gamma_+) \qquad\qquad in \ \ U_\gamma(\Gamma_+)$$

$$F_+(x) \leq -\gamma \qquad\qquad in \ \ \bar{\Omega} \setminus U_\gamma(\Gamma_+)\,.$$

Since it holds that $b \cdot \nabla F_+ > 0$ on Γ_+, we obtain

$$b \cdot \nabla F_+ \geq \beta^2 |\nabla F_+|^2 \qquad\qquad in \ \ U_\gamma(\Gamma_+) \tag{4.3}$$

for sufficiently small γ.
Now let

$$S(x,\varepsilon) = K_4 \varepsilon + K_5 \exp\Big(\frac{F_+(x)}{K_6 \varepsilon}\Big) + \mu(\varepsilon)\,.$$

$S(x,\varepsilon)$ majorizes the boundary conditions of z_ε for sufficiently large K_5.
It holds

$$L_\varepsilon S = K_4 (c - \varepsilon a)\varepsilon + (c-\varepsilon a)\mu(\varepsilon) +$$

$$+ K_5 \exp\Big(\frac{F_+(x)}{K_6 \varepsilon}\Big)\Big[\frac{1}{\varepsilon}\Big(\frac{1}{K_6}\, b \cdot \nabla F_+ - \frac{1}{K_6^2}\sum_{i,j} a_{ij} \frac{\partial F_+}{\partial x_i} \cdot \frac{\partial F_+}{\partial x_j}\Big) - \frac{1}{K_6}\sum a_{ij}\frac{\partial^2 F_+}{\partial x_i \partial x_j} - \frac{1}{K_6}\sum a_i \frac{\partial F_+}{\partial x_i} + c - \varepsilon a\Big].$$

The term in the brackets is certainly positive if

$$b \cdot \nabla F_+ - \frac{1}{K_6}\sum_{i,j} a_{ij}\Big(\frac{\partial F_+}{\partial x_i}\Big)\cdot\Big(\frac{\partial F_+}{\partial x_j}\Big)$$

is positive.
This is the case in $U_\gamma(\Gamma_+)$ because of (4.3) with sufficiently large K_6.
Consequently, for positive $\mu(\varepsilon)$ it holds

$$L_\varepsilon S \geq K_4 (c - \varepsilon a)\varepsilon \qquad\qquad in \ \ U_\gamma(\Gamma_+)$$

and $L_\varepsilon S$ majorizes $|L_\varepsilon z_\varepsilon|$ for the corresponding choice of K_4 in $U_\gamma(\Gamma_+)$.
In $\bar{\Omega} \setminus U_\gamma(\Gamma_+)$ we choose $\mu(\varepsilon)$ in such a way that

$$L_\varepsilon S \geq K_4 (c - \varepsilon a)\varepsilon\,.$$

This can be guaranteed by setting

$$\mu(\varepsilon) = K_5^* \exp\Big(-\frac{\gamma}{K_6 \varepsilon}\Big)\Big[\frac{K_7}{\varepsilon} + K_8\Big]$$

with sufficiently large K_7, K_8.
Summing up we obtain

$$|u_\varepsilon(x)-u(x)| \leq K_4\varepsilon + K_5 \exp\left(\frac{F_+(x)}{K_6\varepsilon}\right) + K_5 \exp\left(-\frac{\gamma}{K_6\varepsilon}\right)\left[\frac{K_7}{\varepsilon} + K_8\right] .$$

From this it follows the assertion by choosing $\gamma = \gamma(\varepsilon)$ for sufficiently small ε in such a way that $\lim_{\varepsilon \to 0} \frac{\varepsilon \ln \varepsilon}{\gamma(\varepsilon)} = 0$. ■

In the case of nonempty Γ_0 we need an additional assumption. If, namely, we have a regular function $F_0(x)$ with the properties

$$F_0(x) = -\,dist(x, \Gamma_+ \cup \Gamma_0) \qquad in\ U_\gamma(\Gamma_+ \cup \Gamma_0)$$
$$F_0(x) \leq -\gamma \qquad in\ \bar{\Omega} \setminus U_\gamma(\Gamma_+ \cup \Gamma_0) ,$$

it holds

$$b \cdot \nabla F_0 \geq 0 \qquad on\ \Gamma_+ \cup \Gamma_0$$

but this is not necessarily satisfied in a neighbourhood of $\Gamma_+ \cup \Gamma_0$.

Theorem 4.2. Let Γ_- be closed in Γ and $c(x) \geq c_0 > 0$ such that $u \in C^2(\bar{\Omega})$.
Furthermore, let a positive γ exist such that the function $F_0(x) = -\,dist(x, \Gamma_+ \cup \Gamma_0)$ satisfies the inequality

$$b \cdot \nabla F_0 \geq 0 \qquad in\ U_\gamma(\Gamma_+ \cup \Gamma_0) .$$

Then there exists a constant K independent of x, ε such that for sufficiently small ε

$$|u_\varepsilon(x) - u(x)| \leq K\varepsilon \qquad in\ \bar{\Omega} \setminus U_{\gamma(\varepsilon)}(\Gamma_+ \cup \Gamma_0)$$

where $\gamma(\varepsilon)$ satisfies $\lim_{\varepsilon \to 0} \frac{\varepsilon^{\frac{1}{2}} \ln \varepsilon}{\gamma(\varepsilon)} = 0$. ■

Proof. The proof is analogous to that of Theorem 4.1.
Setting

$$S(x,\varepsilon) = K_4\varepsilon + K_5 \exp\left(\frac{F_0(x)}{K_6\varepsilon^{\frac{1}{2}}}\right) + \mu(\varepsilon)$$

we obtain

$$L_\varepsilon S = K_4(c-\varepsilon a)\varepsilon + (c-\varepsilon a)\mu(\varepsilon) +$$

$$+ K_5 \exp\left(\frac{F_0(x)}{K_6\varepsilon^{1/2}}\right)\left[\frac{1}{\varepsilon^{1/2}}\left(\frac{1}{K_6} b\cdot\nabla F_0\right) + c - \varepsilon a - \frac{1}{K_6^2}\sum_{i,j} a_{ij}\left(\frac{\partial F_0}{\partial x_i}\right)\left(\frac{\partial F_0}{\partial x_j}\right) - \frac{\varepsilon^{\frac{1}{2}}}{K_6}\left(\sum_{i,j} a_{ij}\frac{\partial^2 F_0}{\partial x_i \partial x_j} + \sum a_i \frac{\partial F_0}{\partial x_i}\right)\right].$$

It holds

$$L_\varepsilon S \geqq K_4(c-\varepsilon a)\varepsilon$$

in $U_\gamma(\Gamma_+ \cup \Gamma_0)$ since $b \cdot \nabla F_0 \geqq 0$ and

$$c-\varepsilon a > \frac{1}{K_6^2}\sum_{i,j} a_{ij}\left(\frac{\partial F_0}{\partial x_i}\right)\left(\frac{\partial F_0}{\partial x_j}\right)$$

for sufficiently large K_6. In $\overline{\Omega} \setminus U_\gamma(\Gamma_+ \cup \Gamma_0)$ we have

$$L_\varepsilon S \geqq K_4(c-\varepsilon a)\varepsilon$$

by choosing

$$\mu(\varepsilon) = K_5^* \exp\left(-\frac{\gamma}{K_6\varepsilon^{1/2}}\right)\left[\frac{K_7}{\varepsilon^{1/2}} + K_8 + \varepsilon^{\frac{1}{2}} K_9\right] \quad . \blacksquare$$

According to Theorem 3.2 it is possible to omit the assumption $c(x) \geqq c_0$ in the Theorems 4.1 and 4.2 if any characteristic through a point of $\overline{\Omega}$ leaves $\overline{\Omega}$ in finite time.

If Γ_- is not closed in Γ, then, by Theorem 3.3 we are motivated to study statements about global approximation in subdomains of the domain of influence of Γ_-.

<u>Theorem 4.3.</u> Let Σ be a connected compact set with $\Sigma \subset \Gamma_-$, let $G(\Sigma)$ denote the domain of influence of Σ and $c(x) > 0$ in $\overline{\Omega}$. Then there exists a constant K independent of x and ε such that

$$|u_\varepsilon(x) - u(x)| \leqq K\varepsilon \qquad in\ G(\Sigma)\setminus\left(\mathcal{U}_\gamma(\Gamma_+) \cup \mathcal{U}_\gamma(I(\overline{\Omega}))\right) .$$

If, furthermore, all characteristics through points of $G(\Sigma)$ leave the domain Ω at points of Γ_+ in finite time, then, more precisely, it holds

$$|u_\varepsilon(x) - u(x)| \leqq K\varepsilon \qquad in\ G(\Sigma)\setminus \mathcal{U}_{\gamma(\varepsilon)}(\Gamma_+)$$

where $\gamma(\varepsilon)$ satisfies $\lim\limits_{\varepsilon\to 0} \dfrac{\varepsilon \ln\varepsilon}{\gamma(\varepsilon)} = 0.$ $\blacksquare$

<u>Proof.</u> As in the proof of Theorem 3.3 we shall argue with compactness. First we estimate $z_\varepsilon = u_\varepsilon - u$ in an open neighbourhood of an arbitrary characteristic $\xi_y(\tau)$, $y \in \Sigma$. For this purpose let $\xi_y(\tau)$ be imbedded in a cylinder of characteristics $\varphi_{y,r}$ (according to LEFSCHETZ (1957)). For fixed $y \in \Sigma$ let

$$B(y,r) = \{\, x \mid |x-y| < r \,\}$$

and

$$\varphi_{y,r} = \{\, \xi_x(\tau) \mid x \in \Sigma \cap B(y,r)\, ,\, 0<\tau<\tau^*(x) \} \ .$$

We choose r to be sufficiently small such that, especially, $B(y,r) \cap \Gamma_- \subset \Sigma$ and set $\tau^*(x) = \tau_+(x)$ in the case of a finite exit time, otherwise $\tau^*(x) < \infty$.
Finally, let

$$\Gamma_+^* = \{ \; \xi_x(\tau^*(x)) \mid x \in \Sigma \cap B(y,r) \}$$

(see figure 4.1).

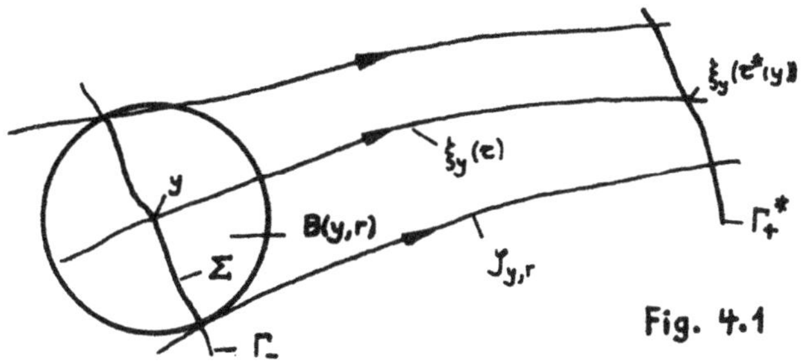

Fig. 4.1

Then it holds that

$$|L_\varepsilon z_\varepsilon| \leq K_1 \varepsilon, \qquad |z_\varepsilon| \leq K_2 \qquad \text{in } \overline{\mathcal{S}}_{y,r},$$
$$z_\varepsilon \big|_\Sigma = 0 .$$

Setting

$$S(x,\varepsilon) = K_3 \varepsilon + K_4 \exp\Big(\frac{F_0(x)}{K_5 \varepsilon^{1/2}}\Big) + K_6 \exp\Big(\frac{F_+(x)}{K_7 \varepsilon}\Big) + \mu(\varepsilon)$$

with regular functions F_0, F_+ we obtain

$$L_\varepsilon S = K_3(c - \varepsilon a)\varepsilon + (c - \varepsilon a)\mu(\varepsilon) + \qquad (*)$$

$$+ K_4 \exp\Big(\frac{F_0(x)}{K_5 \varepsilon^{1/2}}\Big)\Big[\frac{1}{\varepsilon^{1/2}}\Big(\frac{1}{K_5} b \cdot \nabla F_0\Big) + c - \varepsilon a - \frac{1}{K_5^2} \sum a_{ij} \frac{\partial F_0}{\partial x_i}\frac{\partial F_0}{\partial x_j} - \frac{\varepsilon^{1/2}}{K_5}\Big(\sum a_{ij}\frac{\partial^2 F_0}{\partial x_i \partial x_j} + \sum a_i \frac{\partial F_0}{\partial x_i}\Big)\Big] +$$

$$+ K_6 \exp\Big(\frac{F_+(x)}{K_7 \varepsilon}\Big)\Big[\frac{1}{\varepsilon}\Big(\frac{1}{K_7} b \cdot \nabla F_+ - \frac{1}{K_7^2} \sum a_{ij} \frac{\partial F_+}{\partial x_i}\frac{\partial F_+}{\partial x_j}\Big) - \frac{1}{K_7}\Big(\sum a_{ij}\frac{\partial^2 F_+}{\partial x_i \partial x_j} + \sum a_i \frac{\partial F_+}{\partial x_i}\Big) + c - \varepsilon a\Big].$$

By choosing F_0 as a regular solution of

$$b\cdot\nabla F_0 = 0 \qquad in\ \mathcal{S}_{y,r}$$

$$F_0(x) = |x-y| - r \qquad for\ x\in\Sigma\cap\overline{B(y,r)}\setminus B(y,\tfrac{r}{2})$$

$$F_0(x) \leq -\tfrac{r}{2} \qquad for\ x\in\Sigma\cap B(y,\tfrac{r}{2})$$

the first bracket on the right hand side of (*) is non-negative for sufficiently large K_5. The estimation of the second bracket and the choice of $\mu(\varepsilon)$ can be done in the same way as in the proof of Theorem 4.1. Here we set

$$F_+(x) = -\,dist(x,\Gamma_+^*) \qquad in\ U_\gamma(\Gamma_+^*)$$

and

$$F_+(x) \leq -\gamma \qquad in\ \overline{\mathcal{S}}_{y,r}\setminus U_\gamma(\Gamma_+^*)\ .$$

Summing up we obtain

$$L_\varepsilon S \geq K_3(c-\varepsilon a)\varepsilon\ .$$

Since, for sufficiently large K_4 and K_6, z_ε is majorized on the boundary of $\mathcal{S}_{y,r}$ by S, it follows that in $\mathcal{S}_{y,\frac{r}{2}}$

$$|u_\varepsilon(x)-u(x)| \leq K_3\varepsilon + \mu(\varepsilon) + K_6\exp\left(\frac{F_+(x)}{K_7\varepsilon}\right)\ .$$

Due to the compactness $G(\Sigma)\setminus(U_\gamma(\Gamma_+)\cup U_\gamma(I(\overline{\Omega})))$ can be covered by a finite number of cylinders $\mathcal{S}_{y,\frac{r}{2}}$ from which the assertion immediately follows. ∎

Now we consider the parabolic boundary value problem

$$\frac{\partial u_\varepsilon}{\partial t} + L_\varepsilon u_\varepsilon = f \qquad in\ Q$$

$$u_\varepsilon = 0 \qquad on\ S_T \qquad (4.1)_p$$

$$u_\varepsilon = 0 \qquad in\ \Omega\ .$$

Let u be the solution of

$$\frac{\partial u}{\partial t} + L_0 u = f \qquad in\ Q$$

$$(4.2)_p$$

$$u\big|_{\overline{\Omega\cup S_T}} = 0\ .$$

According to Chapter 3 we obtain the simplest case by requiring conditions only at t = 0. Analogously, as in Theorem 4.1, Theorem 4.2, it is possible to prove the assertion of

<u>Theorem 4.4.</u> Let $S_T^- = \phi$.
If, additionally, $S_T^0 = \phi$, then it holds

$$|u_\varepsilon(x,t) - u(x,t)| \leq K\varepsilon \qquad in\ \overline{Q}\setminus U_{\gamma(\varepsilon)}(S_T^+)$$

where $\gamma(\varepsilon)$ satisfies $\lim\limits_{\varepsilon\to 0}\dfrac{\varepsilon \ln\varepsilon}{\gamma(\varepsilon)} = 0.$

If S_T^o is not the empty set and if there is a neighbourhood U_γ of $S_T^+ \cup S_T^o$ with

$$b \cdot \nabla F \geq 0 \qquad in \ \mathcal{U}_\gamma ,$$

then it holds

$$| u_\varepsilon(x,t) - u(x,t) | \leq K\varepsilon \qquad in \ \bar{Q} \setminus \mathcal{U}_{\gamma(\varepsilon)}(S_T^+ \cup S_T^o)$$

provided that $\lim\limits_{\varepsilon \to 0} \dfrac{\varepsilon^{1/2} \ln \varepsilon}{\gamma(\varepsilon)} = 0.$ ■

If S_T^- is not an empty set, but only closed in S_T, then additional difficulties arise from the fact that u is not continuously differentiable (cf. Theorem 3.5).

<u>Theorem 4.5.</u> Let $S_T^- = S_T$.
Then there exists a constant K, independent of x, ε such that for sufficiently small ε

$$| u_\varepsilon(x,t) - u(x,t) | \leq K\varepsilon^{1/2} \qquad in \ \bar{Q} \quad . \quad ■$$

For the proof we need some stronger statements about regularity as stated in Theorem 3.5. Therefore we first prove (let $\bar{x} = (x,t)$)

<u>Lemma 4.1.</u> Let $\| \cdot \|_{p,m}$ be the usual norm in the space $C^{p,\frac{p}{m}}(\bar{Q})$ (cf. 2.1)

(i) For u

$$\sup_{\bar{x} \in Q} ess \ \max\left\{ |u|, \left|\frac{\partial u}{\partial x_1}\right|, \ldots, \left|\frac{\partial u}{\partial x_n}\right| \right\} \leq K_1 \| f \|_{1,m}$$

does exist ($m \geq 1$, arbitrary).

(ii) If, additionally,

$$f(\bar{x}) \geq 0 \qquad for \ all \ \ \bar{x} \in \Gamma_- = \partial\Omega \cap S_T^- ,$$

then it holds that

$$A_o(u,v) = \int_\Omega \sum_{i,j} a_{ij}(\bar{x}) \frac{\partial u}{\partial x_i} \frac{\partial v}{\partial x_j} dx \geq - K_2 \| f \|_{2,2} \int_\Omega v \, dx$$

for all $v \in H_o^1(\Omega)$ with $v \geq 0$.

(iii) If g is a regular function and

$$f = f_1 + \left(\frac{\partial}{\partial t} + L_o \right) g ,$$

we can sharpen the assertion of (ii), namely

$$A_o(u,v) \geq - K_2 (\| f_1 \|_{2,2} + \| g \|_{2,2}) \int_\Omega v \, dx \quad .$$

In the estimates given above the constants K_1, K_2 do not depend on f and g, respectively. ■

Remark. Lemma 4.1 remains also valid in the more general case S_T^- closed in S_T, which has been considered in Theorem 3.5. ∎

Proof of Lemma 4.1. Let $\xi_{\bar{x}}(\tau)$ be the characteristic of the operator $\frac{\partial}{\partial t} + L_0$ through the point $\bar{x} = (x,t)$ and let $t_0(\bar{x}) \leq t$ be the function defined uniquely for all $\bar{x} \in Q$ by the relations

$$\xi_{\bar{x}}(t_0(\bar{x})) \in \Omega \cup S_T \cdot$$

$$\xi_{\bar{x}}(\tau) \in Q \qquad \text{for all } \tau \in (t_0(\bar{x}), t) \ .$$

Then it holds (cf. KAMKE (1965))

$$u(\bar{x}) = \int_{t_0(\bar{x})}^{t} \psi(\tau,\bar{x})\, d\tau$$

$$\psi(\tau,\bar{x}) = exp(-\int_{\tau}^{t} c(\xi_{\bar{x}}(\sigma))\, d\sigma)\, f(\xi_{\bar{x}}(\tau)) \ .$$

$\xi_{\bar{x}}(\tau)$ and $\psi(\tau,\bar{x})$ are regular in x, $t_0(\bar{x})$ is at least continuous for fixed t. The first derivatives of $t_0(\bar{x})$ possess a jump on the manifold Γ_t defined by the relation

$$\xi_{\bar{x}}(t_0(\bar{x})) \in \Gamma \ .$$

But $t_0(\bar{x})$ is regular on the subsets $\bar{\Omega}_t$ and $\bar{\Omega} \setminus \Omega_t$ with

$$\Omega_t = \{\, x \mid \ t_0(\bar{x}) = 0 \} \setminus \Gamma_t \ .$$

Then for $x \in \Omega_0 \setminus \Gamma_t$ it holds

$$\frac{\partial u}{\partial x_i} = \int_{t_0}^{t} \frac{\partial \psi}{\partial x_i}\, d\tau - \frac{\partial t_0}{\partial x_i}\, \psi\Big|_{(t_0,\bar{x})} \quad ,$$

$$\frac{\partial^2 u}{\partial x_i \partial x_j} = \int_{t_0}^{t} \frac{\partial^2 \psi}{\partial x_i \partial x_j}\, d\tau - \Big(\frac{\partial t_0}{\partial x_j}\frac{\partial \psi}{\partial x_i} + \frac{\partial t_0}{\partial x_i}\frac{\partial \psi}{\partial x_j} + \frac{\partial t_0}{\partial x_i}\frac{\partial t_0}{\partial x_j}\frac{\partial \psi}{\partial \tau} + \frac{\partial^2 t_0}{\partial x_i \partial x_j}\psi\Big)\Big|_{(t_0,\bar{x})} \cdot$$

From this it follows by estimating the larger maximum on the sets $\bar{\Omega}_t$ and $\bar{\Omega} \setminus \Omega_t$, respectively,

$$|u| \leq T \max|\psi| \leq K_3 \|f\|_{1,2m} \quad ,$$

$$\Big|\frac{\partial u}{\partial x_i}\Big| \leq K_4 \max\{|\psi| + \Big|\frac{\partial \psi}{\partial x_i}\Big|\} \leq K_5 \|f\|_{1,2m} \ , \tag{*}$$

$$\Big|\frac{\partial^2 u}{\partial x_i \partial x_j}\Big| \leq K_6 \max\{|\psi| + \Big|\frac{\partial \psi}{\partial x_i}\Big| + \Big|\frac{\partial \psi}{\partial x_j}\Big| + \Big|\frac{\partial \psi}{\partial \tau}\Big| + \Big|\frac{\partial^2 \psi}{\partial x_i \partial x_j}\Big|\} \leq K_7 \|f\|_{2,2} \ . \tag{**}$$

For the choice $K_1 = \max(K_3, K_5)$, (i) is proved.

Now we calculate the "jump" of the first derivatives for $x \in \Gamma_t$

$$\varphi_i(\bar{x}) = \lim_{\substack{y \to x \\ y \in \Omega_t}} \frac{\partial u}{\partial x_i}(y,t) - \lim_{\substack{y \to x \\ y \notin \bar{\Omega}_t}} \frac{\partial u}{\partial x_i}(y,t) \ .$$

Since $\int_{t_0}^{t} \frac{\partial \psi}{\partial x_i} d\tau$ is continuous and it holds $t_0(y) = 0$ for $y \in \Omega_t$ it follows that

$$\varphi_i(\bar{x}) = \psi(0,\bar{x}) \frac{\partial t_0}{\partial x_i}(\bar{x}) .$$

Thereby the derivative has to be calculated on the set $\bar{\Omega} \setminus \Omega_t$. With

$$r(\bar{x}) = \Big[\sum_i \Big(\frac{\partial t_0}{\partial x_i}(\bar{x})\Big)^2 \Big]^{\frac{1}{2}}$$

we obtain

$$\varphi_i(\bar{x}) = \psi(0,\bar{x})\, r(\bar{x})\, \nu_i(t)$$

where $\nu(t) = (\nu_i(t))$ denotes the normal on Γ_t directed towards $\Omega_0 \setminus \bar{\Omega}_t$. By means of partial integration it follows

$$A_0(u,v) = -\int_{\Omega \setminus \Gamma_t} v \sum \Big(a_{ij} \frac{\partial^2 u}{\partial x_i \partial x_j} + \frac{\partial a_{ij}}{\partial x_j} \frac{\partial u}{\partial x_i}\Big) dx + \int_{\Gamma_t} v \sum a_{ij}\, \varphi_i\, \nu_j(t)\, d\omega .$$

From

$$\sum a_{ij} \varphi_i \nu_j(t) = f(\xi_{\bar{x}}(0)) \exp\Big(-\int_0^t c(\xi_{\bar{x}}(\sigma)) d\sigma\Big) r(\bar{x}) \sum_{i,j} a_{ij}\, \nu_i(t)\, \nu_j(t) \geq 0$$

and the estimates (*), (**) we finally obtain the assertion (ii). On the same line (iii) can be proved by using the estimate

$$\Big| \frac{\partial^2 u}{\partial x_i \partial x_j} \Big| \leq K_8 \big(\| f_1 \|_{2,2} + \| g \|_{2,2} \big)$$

instead of (**), which can be derived from the characteristics representation

$$u(\bar{x}) - g(\bar{x}) = -g[\xi_{\bar{x}}(t_0(\bar{x}))] \exp\Big(-\int_{t_0(\bar{x})}^{t} c(\xi_{\bar{x}}(\sigma)) d\sigma\Big) + \int_{t_0(\bar{x})}^{t} f_1(\xi_{\bar{x}}(\tau)) \exp\Big(-\int_{\tau}^{t} c(\xi_{\bar{x}}(\sigma)) d\sigma\Big) d\tau . \blacksquare$$

<u>Proof of Theorem 4.5.</u> We first assume that

$$-a(\bar{x}) \geq \frac{1}{2} \sum_{i,j} \frac{\partial^2 a_{ij}}{\partial x_i \partial x_j} - \frac{1}{2} \sum_i \frac{\partial a_i}{\partial x_i} , \qquad c(\bar{x}) \geq \max\Big\{0, \frac{1}{2} \sum_i \frac{\partial b_i}{\partial x_i}\Big\} .$$

Then the generalized maximum principle (Theorem 2.7) can be applied both on the operator $\frac{\partial}{\partial t} + L_\varepsilon$ and on $\frac{\partial}{\partial t} + L_0$. Let

$$z_\varepsilon = u_\varepsilon - u .$$

We construct a barrier function $s = s_1 + s_2$ such that, according to Lemma 4.1, by $s_1 \in H^1(Q)$ we are able to give an upper bound for $(\frac{\partial}{\partial t} + L_\varepsilon)(-s_1 \pm z_\varepsilon)$ by means of a regular distribution, which can be majorized by $(\frac{\partial}{\partial t} + L_\varepsilon)\, s_2$ with $s_2 \in C^{2,1}(\bar{Q})$.

Because of $u_\varepsilon \in C^{2,1}(\bar{Q})$ it is sufficient to consider $w = \pm u + s_1 \in H^1(Q)$. For sufficiently small γ let $F(\bar{x})$ be a regular function with

$$F(\bar{x}) = \begin{cases} dist(\bar{x}, S_T), & \bar{x} \in U_\gamma(S_T) \\ \geq \gamma, & \bar{x} \in \bar{Q} \setminus U_\gamma(S_T) \end{cases}$$

and

$$g(\bar{x}) = K_1 \alpha(F(\bar{x}))$$

with

$$\alpha(\sigma) = \sigma \exp(1 - \tfrac{\sigma}{\lambda}), \qquad \lambda = \lambda(\varepsilon) \leq 1 .$$

(It holds $\alpha'(0) = 1$, $|\alpha^{(l)}(\sigma)| \leq \max\{1,l\}\lambda^{1-l}$ for all $l \geq 0$).
Further, let w be a solution of

$$\frac{\partial w}{\partial t} + L_0 w = \pm f - L_0 g, \qquad x \in Q,$$

$$w = 0, \qquad x \in \Omega \cup S_T .$$

For $\bar{x} \in \partial\Omega$ and sufficiently large K_1 it holds that

$$\pm f(\bar{x}) - L_0 g(\bar{x}) = \pm f(\bar{x}) - K_1 b \cdot \nu \geq 0 .$$

According to Lemma 4.1, for arbitrary $v \in H_0^1(\Omega)$, it follows that

$$\int_\Omega \Big[\sum a_{ij} \frac{\partial w}{\partial x_i}\frac{\partial v}{\partial x_j} + v\Big(\sum_i(-a_i + \sum_j \frac{\partial a_{ij}}{\partial x_j})\frac{\partial w}{\partial x_i} - aw\Big)\Big] dx \geq -K_2(\|f\|_{2,2} + \|g\|_{2,2})\int_\Omega v\, dx$$

$$\geq -\frac{K_3}{\lambda}\int_\Omega v\, dx ,$$

consequently

$$\Big(\frac{\partial}{\partial t} + L_\varepsilon\Big)(-w) \leq K_3 \frac{\varepsilon}{\lambda} \mp f + L_0 g$$

and

$$\Big(\frac{\partial}{\partial t} + L_\varepsilon\Big)(-s_1 \pm z_\varepsilon) \leq K_3 \frac{\varepsilon}{\lambda} + L_0 g ,$$

respectively. Since $|L_1 g(\bar{x})| \leq K_4 \lambda^{-1}$

$$s(\bar{x}) = s_1(\bar{x}) + g(\bar{x}) + \frac{\varepsilon}{\lambda}(K_3 + K_4)(1 + T)$$

satisfies the relation

$$\Big(\frac{\partial}{\partial t} + L_\varepsilon\Big)(-s \pm z_\varepsilon) \leq 0 .$$

According to Theorem 2.7 from

$$-s(\bar{x}) \pm z_\varepsilon(\bar{x}) = -g(\bar{x}) - \frac{\varepsilon}{\lambda}(K_3 + K_4)(1+T) \leq 0$$

it follows that

$$|z_\varepsilon(\bar{x})| \leq s(\bar{x}) .$$

s_1 satisfies the conditions

$$\left(\frac{\partial}{\partial t} + L_0\right)(s_1 + g - K_1\lambda) = -K_1\lambda c \leq 0\,, \qquad \bar{x} \in Q$$

$$s_1 + g - K_1\lambda = g - K_1\lambda \leq 0\,, \qquad \bar{x} \in \Omega \cup S_T$$

in correspondence to the construction of w. From this it follows ($g \geq 0$)

$$s_1(\bar{x}) \leq K_1\lambda\,,$$

$$s(\bar{x}) \leq 2K_1\lambda + \frac{\varepsilon}{\lambda}(K_3 + K_4)(1+T)\,,$$

consequently

$$|u_\varepsilon(\bar{x}) - u(x)| \leq s(\bar{x}) \leq K\varepsilon^{\frac{1}{2}}\,.$$

If the assumptions made at the beginning for $a(\bar{x})$ and $c(\bar{x})$ are not fulfilled, then by means of

$$u_\varepsilon(\bar{x}) = \exp\big(K_5(1+\varepsilon)t\big)\, v_\varepsilon(\bar{x})\,, \qquad u(\bar{x}) = \exp\big(K_5(1+\varepsilon)t\big)\, v(\bar{x})$$

the problem is transformed into

$$\left(\frac{\partial}{\partial t} + L_\varepsilon + K_5(1+\varepsilon)\right)v_\varepsilon = \exp\big(-K_5(1+\varepsilon)t\big)f\,,$$

$$\left(\frac{\partial}{\partial t} + L_0 + K_5(1+\varepsilon)\right)v = \exp\big(-K_5(1+\varepsilon)t\big)f$$

and K_5 is chosen sufficiently large . ■

If S_T^- is closed in S_T but differs from S_T, then we can obtain statements concerning global approximations by combining the Theorems 4.4 and 4.5.

In case of S_T^- is not closed in S_T, we have to combine the statements of the Theorems 4.3 (this Theorem is analogously true in the parabolic case), 4.4 and 4.5.

4.2. Asymptotic solutions for equations of second order in the simplest case

In Section 4.1 we have seen that the solution of the global problem is an asymptotic solution, but only in subdomains. According to the Theorems 4.2 and 4.4 for the construction of asymptotic solutions it is necessary to construct local correctors in the neighbourhood of Γ_+ and Γ_0 and S_T^+ and S_T^0, respectively.
Analogous to the Theorems 4.2 and 4.4 we shall assume in the present section that in the elliptic case Γ_- is closed in Γ and $c(x) \geq c_0$ sufficiently large and that in the parabolic case S_T^- is closed in S_T. But, now we have to sharpen this assumptions.
On the one hand it is difficult to construct local correctors if $\bar{\Gamma}_+ \cap \Gamma_0$ (and $\bar{S}_T^+ \cap S_T^0$, respectively) is not the empty set (see Section 4.4), on the other hand we shall now construct approximations of arbitrary accuracy for which sufficiently regular solutions of the global problem are necessary. Therefore, in the present section we shall assume that

(H1) $\bar{\Gamma}_\pm \cap \Gamma_0 = \phi$ and $\bar{S}_T^\pm \cap S_T^0 = \phi$, respectively

and in the elliptic case additionally,

(H2)$_e$ $\inf_{x\in\Omega} c(x) \geq c_0$, with sufficiently large c_0

are fulfilled. In the case of a simply connected domain Ω from (H1) it follows $\Gamma = \Gamma_+$ or $\Gamma = \Gamma_-$ or $\Gamma = \Gamma_0$. These three cases will be discussed first. The results can be extended to the case of multiply connected domains Ω if for each connected component it holds that $\partial\Omega_i = \Gamma_+$ or $\partial\Omega_i = \Gamma_-$ or $\partial\Omega_i = \Gamma_0$. These results are stated in Section 4.2.4.

4.2.1. Uniform asymptotic solutions in the case $\Gamma = \Gamma_+$

Let us start with the study of the elliptic case. The global formal solution

$$u^N(x,\varepsilon) = \sum_{\mu=0}^{N} \varepsilon^\mu u_\mu(x)$$

satisfies the equations

$$L_0 u_0 = f, \qquad L_0 u_\mu = -L_1 u_{\mu-1}, \qquad \mu = 1,\dots,N .$$

Because of (H2)$_e$ and $\partial\Omega = \Gamma_+$ the u_μ are uniquely determined and regular according to Theorem 3.1. In general, the global formal solution does not fulfil the boundary condition on $\partial\Omega$. Therefore, for the construction of formal solutions, local correctors in the neighbourhood of the boundary are needed.

Let $(\rho,\varphi) = (\rho, \varphi_1, \dots, \varphi_{n-1})$ be local coordinates in a neighbourhood of $\partial\Omega$ such that $\rho(x) = \text{dist}\,(x, \partial\Omega)$ and $0 < \rho < \rho_0$ corresponds to a strip parallel to $\partial\Omega$. L_ε transforms into

$$\tilde{L}_\varepsilon \equiv \varepsilon \sum_{l+|\beta| \le 2} A_{l\beta}(\rho,\varphi) D_\rho^l D_\varphi^\beta + B_0(\rho,\varphi) D_\rho + \sum_{\mu=1}^{n-1} B_\mu(\rho,\varphi) \frac{\partial}{\partial \varphi_\mu} + C(\rho,\varphi) \quad .$$

Because of $B_0(0,\varphi) = b \cdot \nabla \rho \big|_{\rho=0} < 0$, we set $\zeta = \frac{\rho}{\varepsilon}$ for the local variable and

$$v^N(\zeta, \varphi, \varepsilon) = \sum_{\mu=0}^{N} \varepsilon^\mu v_\mu(\zeta, \varphi)$$

for the local corrector. We obtain the equations for the v_μ $(\mu = 0, 1, \dots, N)$

$$L v_0 = 0 \ , \qquad L v_\mu = F_\mu(v_0, \dots, v_{\mu-1}) \ , \qquad \mu = 1, \dots, N$$

with

$$L v \equiv A_{2,0}(0,\varphi) \frac{\partial^2 v}{\partial \zeta^2} + B_0(0,\varphi) \frac{\partial v}{\partial \zeta} \quad .$$

The F_μ are multilinear differential operators, the coefficients of which are polynomials in ζ. To satisfy the boundary conditions we set

$$v_\mu \big|_{\zeta=0} = - u_\mu \big|_{\rho=0} \quad .$$

Because of the uniform ellipticity of L_1 it holds $A_{2,0}(0,\varphi) < 0$ and from the above we have $B_0(0,\varphi) < 0$. Consequently, according to Appendix A 1 there are uniquely determined solutions of classical boundary layer type, especially

$$v_0(\zeta,\varphi) = - u_0 \big|_{\Gamma_+} \exp\left(- \frac{B_0(0,\varphi)}{A_{2,0}(0,\varphi)} \zeta\right) \quad .$$

Let $\psi \in C^\infty(\overline{\Omega})$ be a cut-off function with $0 \le \psi(x) \le 1$ and

$$\psi(x) = \begin{cases} 1 & \text{for} \quad 0 \le \rho(x) \le \frac{\rho_0}{3} \\ 0 & \text{for} \quad \rho(x) \ge \frac{2}{3}\rho_0 \end{cases} \quad .$$

Setting

$$\Phi_\varepsilon^N(x) = u^N(x,\varepsilon) + \psi(x)\, v^N(\zeta, \varphi, \varepsilon)$$

we have

<u>Theorem 4.6</u>$_e$. Under the assumptions (H1), $(H2)_e$ $\Phi_\varepsilon^N(x)$ is a uniform asymptotic solution of $(4.1)_e$ in the case $\partial\Omega = \Gamma_+$. It holds

$$\| u_\varepsilon - \Phi_\varepsilon^N \| \le K \varepsilon^{N+1} \quad . \qquad \blacksquare$$

Proof. We consider the difference

$$z_\varepsilon = u_\varepsilon - \Phi_\varepsilon^{N+1} .$$

According to the construction, z_ε satisfies the boundary value problem

$$L_\varepsilon z_\varepsilon = r_\varepsilon \quad in\ \Omega , \qquad z_\varepsilon = 0 \quad on\ \partial\Omega .$$

In view of an application of the a-priori estimate (Theorem 2.10 (i)) deduced in Chapter 2

$$\| z_\varepsilon \| \leq K \| L_\varepsilon z_\varepsilon \|$$

we first estimate $\| r_\varepsilon \|$. It holds the representation

$$r_\varepsilon = -\varepsilon^{N+2} L_1 u_{N+1} - \psi L_\varepsilon v^{N+1} + \sum_{\substack{1 \leq |\beta| \leq 2 \\ |\gamma| \leq 1}} d^\varepsilon_{\beta\gamma}(x)\, D^\beta \psi\, D^\gamma v^{N+1}$$

with uniformly bounded $d^\varepsilon_{\beta\gamma}$. According to the construction of the local corrector v^{N+1} the second term is of the order ε^{N+1} . The third term is different from zero only in $\frac{\varrho_0}{3} \leq \varrho(x) \leq \frac{2\varrho_0}{3}$. However, in this domain all v_μ and their derivatives decrease exponentially such that we finally obtain

$$\| r_\varepsilon \| \leq K \varepsilon^{N+1} .$$

The triangle inequality concludes the proof of Theorem 4.6_e. ■

Remark. The a-priori estimate

$$| z_\varepsilon |_{2+\alpha} \leq K \varepsilon^{1+\alpha} \| L_\varepsilon z_\varepsilon \|_\alpha + K \| L_\varepsilon z_\varepsilon \|$$

yields even

$$| u_\varepsilon - \Phi_\varepsilon^N |_{2+\alpha} \leq K \varepsilon^{N+1} .$$

From this we obtain

$$\lim_{\varepsilon \to 0} \| \varepsilon^l D^\beta u_\varepsilon - \frac{\partial^l v_0}{\partial \zeta^l} \| = 0 , \qquad |\beta| = l, \quad l = 1,2,\ldots$$

that means we also have uniform approximations of the derivatives of u_ε. ■

In the parabolic case the procedure is analogous. The global formal solution

$$u^N(x,t,\varepsilon) = \sum_{\mu=0}^{N} \varepsilon^\mu u_\mu(x,t)$$

satisfies

$$\begin{aligned} &\frac{\partial u_0}{\partial t} + L_0 u_0 = f , && \frac{\partial u_\mu}{\partial t} + L_0 u_\mu = -L_1 u_{\mu-1} , \quad \mu = 1,\ldots,N, \\ &u_0|_{t=0} = 0 , && u_\mu|_{t=0} = 0 . \end{aligned} \tag{4.4}$$

For the local corrector

$$v^N(\zeta,\varphi,t,\varepsilon) = \sum_{\mu=0}^{N} \varepsilon^\mu v_\mu(\zeta,\varphi,t)$$

we now obtain the relations

$$L v_0 = 0$$

$$L v_\mu = F_\mu(v_0,\dots,v_{\mu-1}) + \frac{\partial v_{\mu-1}}{\partial t}, \qquad \mu = 1,\dots,N$$

$$v_\mu\big|_{\zeta=0} = -u_\mu\big|_{\varrho=0}$$

where

$$L v \equiv A_{2,0}(0,\varphi,t)\frac{\partial^2 v}{\partial \zeta^2} + B_0(0,\varphi,t)\frac{\partial v}{\partial \zeta}$$

and the F_μ are multilinear differential operators, the coefficients of which are polynomials in ζ. Because of $A_{20} < 0$ and $B_0 < 0$, there are uniquely determined solutions v_μ, $\mu = 0, \dots, N$ of classical boundary layer type (see Appendix A 1). Contrary to the elliptic case we have to study the behaviour near to $t = 0$. From the representation

$$v_0(\zeta,\varphi,t) = -u_0\big|_{\varrho=0} \exp\Big(-\frac{B_0(0,\varphi,t)}{A_{2,0}(0,\varphi,t)}\zeta\Big)$$

it directly follows that $v_0(\zeta,\varphi,0) = 0$. Because $F_1(v_0)$ does not contain derivatives of v_0 with respect to t, we have $F_1(v_0)\big|_{t=0} = 0$. From the compatibility condition

(H3) $\quad f(x,0) = 0 \qquad$ on $\partial\Omega$

and from (4.4) we obtain

$$\frac{\partial v_0}{\partial t}(\zeta,\varphi,0) = -\frac{\partial u_0}{\partial t}\Big|_{\substack{t=0\\ \varrho=0}} \exp\Big(-\frac{B_0(0,\varphi,0)}{A_{2,0}(0,\varphi,0)}\zeta\Big) = 0 \quad .$$

We notice that L does not contain derivatives with respect to t such that $V_1(\zeta,\varphi,0)$ satisfies the boundary layer problem

$$L v_1\big|_{t=0} = 0 \;, \qquad v_1\big|_{\substack{t=0\\ \zeta=0}} = 0 \quad .$$

From this it finally follows that $v_1(\zeta,\varphi,0) = 0$.
Without additional assumptions concerning f all following functions v_μ, $\mu = 2, \dots, N$ do not satisfy the relation $v_\mu\big|_{t=0} = 0$. This can be seen as follows. As above we have $F_2(v_0, v_1)\big|_{t=0} = 0$ but $\frac{\partial v_1}{\partial t}\Big|_{t=0}$ is zero if and only if $\frac{\partial^2 u_0}{\partial t^2}\Big|_{\varrho=0} = 0$.
Therefore we now assume the sharpened compatibility condition of order N

($\widetilde{\text{H}}$3) $\quad \frac{\partial^i f}{\partial t^i}(x,0) = 0 \qquad$ in Ω, $\qquad i = 0,1,\dots,N$.

Lemma 4.2. Let the sharpened compatibility condition $(\widetilde{H3})$ of order N be fulfilled.
Then it holds that

$$\frac{\partial^i u_\mu}{\partial t^i}(x,0) = 0 \qquad \text{in } \Omega \qquad \text{for} \quad i = 0,1,\dots,N+1, \quad \mu \geq 0$$

$$v_\mu(\zeta,\varphi,0) = 0 \qquad \text{for} \quad \mu = 0,1,\dots,N+1 . \qquad \blacksquare$$

Proof. The first assertion is obvious in the case i = 0. Let the assertion be valid for all j with j ≤ i. From (4.4) it follows that

$$\left.\frac{\partial^{i+1} u_\mu}{\partial t^{i+1}}\right|_{t=0} = \begin{cases} \left.\frac{\partial^i f}{\partial t^i}\right|_{t=0} = 0 , & \mu = 0 \\ -L_1 \left.\frac{\partial^i u_{\mu-1}}{\partial t^i}\right|_{t=0} = 0 , & \mu \geq 1 . \end{cases}$$

Above, it was shown that the second assertion, in the cases $\mu = 0$ and $\mu = 1$, is already valid under the weaker assumption (H3). Because the term

$$\frac{\partial v_{\mu-1}}{\partial t} + F_\mu(v_0,\dots,v_{\mu-1})$$

linearly depends on $\left. D_\varphi^\beta D_t^i u_\varkappa \right|_{\substack{t=0\\ \varrho=0}}$, $\varkappa = 0,1,\dots,\mu-1$; $i = 0,1,\dots,\mu-\varkappa-1$,

it holds

$$M_0 v_\varkappa\big|_{t=0} = 0 , \qquad v_\varkappa\big|_{\substack{t=0\\ \zeta=0}} = 0 , \qquad \varkappa = 0,1,\dots,N+1$$

from which it finally follows the second assertion of Lemma 4.2 ∎

Setting, as in the elliptic case with a cut-off function ψ,

$$\phi_\varepsilon^N(x,t) = u^N(x,t,\varepsilon) + \psi(x)\, v^N(\zeta,\varphi,t,\varepsilon)$$

we obtain

Theorem 4.6 $_p$. Under the assumptions (H1), (H3) and $(\widetilde{H3})$, respectively, $\phi_\varepsilon^N(x,t)$ is a uniform asymptotic solution of $(4.1)_p$ in the case $\partial\Omega = \Gamma_+$. It holds

$$\| u_\varepsilon - \phi_\varepsilon^0 \| \leq K\varepsilon \qquad \text{for (H3)},$$

$$\| u_\varepsilon - \phi_\varepsilon^N \| \leq K\varepsilon^{N+1} \qquad \text{for } (\widetilde{H3}). \qquad \blacksquare$$

Proof. Analogous to the proof of Theorem 4.6$_e$ the difference $z_\varepsilon = u_\varepsilon - \phi_\varepsilon^{N+1}$ satisfies the boundary value problem

$$\frac{\partial z_\varepsilon}{\partial t} + L_\varepsilon z_\varepsilon = r_\varepsilon \qquad \text{in } Q$$

$$z_\varepsilon = 0 \qquad \text{on } S_T$$

$$z_\varepsilon\big|_{t=0} = 0$$

where $\|r_\varepsilon\| \leq K\varepsilon^{N+1}$. Applying the a-priori estimate (Theorem 2.12) deduced in Chapter 2 and the triangle inequality we conclude the proof of Theorem 4.6$_p$ ∎

4.2.2. Uniform asymptotic solutions in the case $\Gamma = \Gamma_-$

According to Theorem 4.1 in the elliptic case it already holds that

$$\| u_\varepsilon(x) - u(x) \| \leq K\varepsilon .$$

Setting

$$u^N(x,\varepsilon) = \sum_{\mu=0}^{N} \varepsilon^\mu u_\mu(x)$$

with

$$L_0 u_0 = f$$

$$L_0 u_\mu = -L_1 u_{\mu-1}, \qquad \mu = 1,\dots,N,$$

$$u_\mu = 0 \quad \text{on } \Gamma_-, \qquad \mu = 0,\dots,N,$$

we obtain

$$\| u_\varepsilon(x) - u^N(x,\varepsilon) \| \leq K\varepsilon^{N+1}$$

as in Theorem 4.1 or in Section 4.2.1. Also in this case it holds

$$| u_\varepsilon - u^N |_{2+\alpha} \leq K\varepsilon^{N+1}.$$

According to Theorem 4.5 we have

$$\| u_\varepsilon(x,t) - u(x,t) \| \leq K\varepsilon^{1/2}$$

in the parabolic case. The construction of approximations of arbitrary exactness requires tedious calculations. Assuming the sharpened compatibility condition ($\tilde{H}3$), we have

$$\| u_\varepsilon - u^N \| \leq K\varepsilon^{N+1}$$

as in the elliptic case.

4.2.3. Uniform asymptotic solutions in the case $\Gamma = \Gamma_0$

The procedure is analogous to the case $\partial\Omega = \Gamma_+$. However, in difference to this case several types of boundary layer equations depending on the behaviour of coefficients appear.
The global formal solution of $(4.1)_\varepsilon$

$$u^N(x,\varepsilon) = \sum_{\mu=0}^{N} \varepsilon^{\mu} u_\mu(x)$$

again satisfies

$$L_0 u_0 = f ,$$

$$L_0 u_\mu = - L_1 u_{\mu-1} , \qquad \mu = 1,\dots,N .$$

Because of $\partial\Omega = \Gamma_0$ and $(H2)_e$, the solutions u_μ are uniquely determined and regular according to Theorem 3.1. We now set $\zeta = \frac{\varrho}{\sqrt{\varepsilon}}$ for the local variable and

$$v^N(\zeta,\varphi,\varepsilon) = \sum_{\mu=0}^{2N} \varepsilon^{\frac{\mu}{2}} v_\mu(\zeta,\varphi)$$

for the local corrector. Then we obtain equations of the form

$$\overset{\circ}{L} v_\mu \equiv A_{2,0}(0,\varphi)\frac{\partial^2 v_\mu}{\partial\zeta^2} + \frac{\partial B_0}{\partial\varrho}(0,\varphi)\,\zeta\,\frac{\partial v_\mu}{\partial\zeta} + \sum_{\varkappa=1}^{n-1} B_\varkappa(0,\varphi)\frac{\partial v_\mu}{\partial\varphi_\varkappa} + C(0,\varphi)v_\mu = F_\mu \tag{4.5}$$

$$F_0 = 0 , \qquad F_\mu = F_\mu(v_0,\dots,v_{\mu-1}) , \quad \mu = 1,\dots,N$$

with multilinear differential operators F_μ, the coefficients of which are polynomials in ζ. Now the problem is the following: Are there solutions of boundary layer type of problem (4.5) with the boundary condition

$$v_\mu\Big|_{\zeta=0} = \begin{cases} -u_\mu\big|_{\varrho=0} & \text{for } \mu = 2\varkappa \\ 0 & \text{for } \mu = 2\varkappa+1 , \ \varkappa = 0,1,\dots, \end{cases} \quad ?$$

We have the simplest case if $\frac{\partial B_0}{\partial\varrho}(0,\varphi) \equiv 0$ and $B_\varkappa(0,\varphi) \equiv 0$ $(\varkappa=1,\dots,n-1)$. Then v_0 is a function of classical boundary layer type and the treatment of the inhomogenous problems is possible without any difficulties. As in the case $\partial\Omega = \Gamma_+$ we can prove

<u>Theorem 4.7$_e$.</u> Under the assumptions (H1), $(H2)_e$ and $\frac{\partial B_0}{\partial\varrho}(0,\varphi) \equiv 0$, $B_\varkappa(0,\varphi) \equiv 0$ for $\varkappa = 1,\dots,n-1$

$$\phi_\varepsilon^N(x) = u^N(x,\varepsilon) + \psi(x)\, v^N(\zeta,\varphi,\varepsilon)$$

is a uniform asymptotic solution of $(4.1)_\varepsilon$ in the case $\partial\Omega = \Gamma_0$.
It holds

$$\| u_\varepsilon - \phi_\varepsilon^N \| \leq K \varepsilon^{N+\frac{1}{2}} \qquad \blacksquare$$

We are confronted with considerable difficulties if the assumptions of Theorem 4.7 are not fulfilled. So, for $\frac{\partial B_0}{\partial \varrho}(0,\varphi) \neq 0$ solutions of classical boundary layer type do not always exist. Furthermore, the treatment of the corresponding inhomogenous problems becomes difficult. Therefore we assume $N = 0$ and

$(H4)_e$ Let one of the following conditions be fulfilled.

(1) $\frac{\partial B_0}{\partial \varrho}(0,\varphi) \neq 0$, $B_\varkappa(0,\varphi) \equiv 0$, $\varkappa = 1,\dots,n-1$,

if $\frac{\partial B_0}{\partial \varrho}(0,\varphi) > 0$ let additionally $C(0,\varphi) > \frac{\partial B_0}{\partial \varrho}(0,\varphi)$,

(2) $\frac{\partial B_0}{\partial \varrho}(0,\varphi) \equiv 0$, $B_\varkappa(0,\varphi) \equiv 0$ for $\varkappa \neq k$, $\varkappa = 1,\dots,n-1$, $B_k(0,\varphi) \neq 0$,

(3) $\frac{\partial B_0}{\partial \varrho}(0,\varphi) \neq 0$, $B_\varkappa(0,\varphi) \equiv 0$ for $\varkappa \neq k$, $\varkappa = 1,\dots,n-1$, $B_k(0,\varphi) \neq 0$,

$\frac{\partial B_0}{\partial \varrho}(0,\varphi) / A_{2,0}(0,\varphi)$ and $C(0,\varphi) / A_{2,0}(0,\varphi)$ independent on φ_k,

if $\frac{\partial B_0}{\partial \varrho}(0,\varphi) > 0$ let additionally $C(0,\varphi) > \frac{\partial B_0}{\partial \varrho}(0,\varphi)$.

Then it holds

<u>Theorem 4.8e.</u> Under the assumptions (H1), $(H2)_e$, $(H4)_e$

$$\phi_\varepsilon(x) = u_0(x) + \psi(x)\, V_0(\zeta,\varphi)$$

is a uniform asymptotic solution of $(4.1)_e$ in the case $\partial\Omega = \Gamma_0$.
It holds

$$\| u_\varepsilon - \phi_\varepsilon \| \leq K \varepsilon^{1/2}. \qquad \blacksquare$$

<u>Proof.</u> Because of the a-priori estimates of Chapter 2 the assertion is proved if, for the difference $z_\varepsilon = u_\varepsilon - \phi_\varepsilon$, an estimate of the form $\| L_\varepsilon z_\varepsilon \| \leq K \varepsilon^{1/2}$ can be shown. We obtain this estimate in the same way as in the case $\partial\Omega = \Gamma_+$ if V_0 is a function of classical boundary layer type.
For (1) from $(H4)_e$ we divide (4.5) by $-A_{2,0}(0,\varphi) > 0$ and obtain the boundary layer problem A 1.3. In case $\frac{\partial B_0}{\partial \varrho}(0,\varphi) < 0$, (4.5) has a solution V_0 of classical boundary layer type and it easily follows that $\| L_\varepsilon z_\varepsilon \| \leq K \varepsilon^{1/2}$. To obtain this estimate also in the case $\frac{\partial B_0}{\partial \varrho}(0,\varphi) > 0$, we have to guarantee that $\| \psi \tilde{L}_\varepsilon V_0 \| \leq K \varepsilon^{1/2}$ and $\| D^\beta \psi D^\gamma V_0 \| \leq K \varepsilon^{1/2}$ for $1 \leq |\beta| \leq 2$, $|\gamma| \leq 1$. For this purpose the additional requirement for C in $(H4)_e$ is sufficient.

Let (2) from $(H4)_e$ be fulfilled. Dividing the equation (4.5) by $C > 0$ and normalizing the variable φ_k, we have to study the boundary layer problem A 3.1. Because V_0 is a function of classical boundary layer type, we have $\| L_\varepsilon z_\varepsilon \| \leq K \varepsilon^{1/2}$.

Under the special assumptions (3) from $(H4)_e$ we investigate the boundary layer equation (4.5) in A 3.3. In the case $\frac{\partial B_0}{\partial \varrho}(0,\varphi) < 0$ we again obtain a solution of classical boundary layer type and for $\frac{\partial B_0}{\partial \varrho}(0,\varphi) > 0$ the additional requirement for C guarantees the validity of the estimate $\| L_\varepsilon z_\varepsilon \| \leq K \varepsilon^{1/2}$ ∎

In the parabolic case the procedure is analogous. Let

$$u^N(x,t,\varepsilon) = \sum_{\mu=0}^{N} \varepsilon^\mu u_\mu(x,t)$$

and let the u_μ be the uniquely determined regular solutions of

$$\frac{\partial u_0}{\partial t} + L_0 u_0 = f$$

$$\frac{\partial u_\mu}{\partial t} + L_0 u_\mu = -L_1 u_{\mu-1}\,, \qquad \mu = 1,\dots,N$$

$$u_\mu\big|_{t=0} = 0\,, \qquad \mu = 0,1,\dots,N\,.$$

For the local corrector

$$V^N(\zeta,\varphi,\varepsilon) = \sum_{\mu=0}^{2N} \varepsilon^{\mu/2} V_\mu(\zeta,\varphi)$$

we obtain

$$\frac{\partial V_\mu}{\partial t} + L V_\mu = F_\mu$$

$$V_\mu\big|_{t=0} = 0 \tag{4.6}$$

$$V_\mu\big|_{\zeta=0} = \begin{cases} 0 & \text{for } \mu = 2\varkappa+1 \\ -u_\mu\big|_{\varrho=0} & \text{for } \mu = 2\varkappa\,, \qquad \varkappa = 0,1,\dots,N\,. \end{cases}$$

First let again $\frac{\partial B_0}{\partial \varrho}(0,\varphi,t) \equiv 0$ and $B_\varkappa(0,\varphi,t) \equiv 0$ for $\varkappa = 1,\dots,n-1$. Transformations of the form $V = V^* \exp(-\int_0^t C(0,\varphi,\tau)d\tau)$ and $t^* = -\int_0^t A_{2,0}(0,\varphi,\tau)d\tau$ guarantee that without restriction of generality we can set $C \equiv 0$ and $A_{2,0} \equiv -1$.

Then we obtain the boundary layer problems studied in A 2.1. The V_μ are solutions of classical boundary layer type if $\frac{\partial^k u_\mu}{\partial t^k}\Big|_{\substack{\varrho=0\\ t=0}} = 0$ for sufficiently large K.

According to Lemma 4.2 this holds under the sharpened compatibility conditions ($\tilde{H}3$). So, we have

Theorem 4.7_p. Under the assumptions (H1), ($\tilde{H}3$), $\frac{\partial B_0}{\partial \varrho}(0,\varphi,t) \equiv 0$ and $B_{\varkappa}(0,\varphi,t) \equiv 0$ for $\varkappa = 1,\dots,n-1$

$$\phi_\varepsilon^N(x,t) = u^N(x,t,\varepsilon) + \psi(x)\, v^N(\zeta,\varphi,t,\varepsilon)$$

is a uniform asymptotic solution of $(4.1)_p$ in the case $\partial\Omega = \Gamma_0$. It holds that

$$\| u_\varepsilon - \phi_\varepsilon^N \| \leq K\, \varepsilon^{N+\frac{1}{2}} \,. \quad \blacksquare$$

Remark. Renouncing ($\tilde{H}3$), from A 2.1 we can see that at least v_0 is a function of classical boundary layer type such that

$$\| u_\varepsilon - u_0 - \psi v_0 \| \leq K\, \varepsilon^{1/2} \quad \blacksquare$$

Now, let

$(H4)_p$ $B_{\varkappa}(0,\varphi,t) \equiv 0$, $\varkappa = 1,\dots,n-1$, $\varkappa \neq k$, $B_k(0,\varphi,t) \neq 0$, $A_{2,0}(0,\varphi,t)$ and $C(0,\varphi,t)$ independent of φ_k

be fulfilled. As above we can assume $A_{2,0} \equiv -1$ and $C \equiv 0$ without restriction of generality. The problems resulting from (4.6) are investigated in A 2.3. v_0 is a function of classical boundary layer type and we obtain

Theorem 4.8_p. Under the assumptions (H1), ($\tilde{H}3$), $(H4)_p$

$$\phi_\varepsilon(x,t) = u_0(x,t) + \psi(x)\, v_0(\zeta,\varphi,t)$$

is a uniform asymptotic solution of $(4.1)_p$ in the case $\partial\Omega = \Gamma_0$. It holds

$$\| u_\varepsilon - \phi_\varepsilon \| \leq K\, \varepsilon^{1/2} \quad \blacksquare$$

4.2.4. Remarks on multiply connected domains

In the first three sections we have assumed that $\partial\Omega$ is connected. Under the assumption (H_1) $\Gamma_\pm \cap \Gamma_0 = \phi$, from this it followed that $\partial\Omega = \Gamma_+$, $\partial\Omega = \Gamma_-$ or $\partial\Omega = \Gamma_0$. Now we allow $\partial\Omega$ to consist of more than one connected components $\partial\Omega_i$, $i = 1,\dots,p$. Then it holds $\partial\Omega_i = \Gamma_+$, $\partial\Omega_i = \Gamma_-$ or $\partial\Omega_i = \Gamma_0$ on each connected component. Let ψ_i be cut-off functions and let $v_0^{(i)}$ be local correctors in a neighbourhood of $\partial\Omega_i$, $i = 1,\dots,p_1$ (exactly for $1 \leq i \leq p_1$ let $\partial\Omega_i = \Gamma_+$ or $\partial\Omega_i = \Gamma_0$). Then, according to the previous sections, under corresponding assumptions it follows that

$$\phi_\varepsilon = u_0 + \sum_{i=1}^{p_1} \psi_i\, v_0^{(i)}$$

is a uniform asymptotic solution of (4.1). Thereby it holds that

$$\| u_\varepsilon - \phi_\varepsilon \| \leq K \varepsilon^\sigma$$

with

$$\sigma = \tfrac{1}{2} \quad if \quad \Gamma_0 \neq \phi \quad and \quad \sigma = 1 \quad if \quad \Gamma_0 = \phi \,.$$

4.3. Asymptotic solutions for equations of arbitrary order

In this section we aim at extending the results of the previous section to equations of arbitrary order.

We shall consider the elliptic boundary value problem

$$\begin{aligned} L_\varepsilon u_\varepsilon \equiv \varepsilon L_1 u_\varepsilon + L_0 u_\varepsilon &= f \qquad in \ \Omega \\ \frac{\partial^s u_\varepsilon}{\partial \nu^s} &= 0 \qquad on \ \partial\Omega, \quad s = 0,1,\ldots,m-1 \end{aligned} \tag{4.7$_e$}$$

where L_1 is uniformly elliptic of order 2m and

$$L_0 u \equiv \sum_{i=1}^{n} b_i \frac{\partial u}{\partial x_i} + c(x) u \qquad ,$$

as well as the parabolic boundary value problem

$$\begin{aligned} \frac{\partial u_\varepsilon}{\partial t} + L_\varepsilon u_\varepsilon &= f \qquad in \ Q \\ \frac{\partial^s u_\varepsilon}{\partial \nu^s} &= 0 \qquad on \ S_T, \quad s = 0,1,\ldots,m-1 \\ u_\varepsilon &= 0 \qquad in \ \Omega \,. \end{aligned} \tag{4.7$_p$}$$

Analogous to 4.2 we shall assume

(H1) $\quad \bar{\Gamma}_\pm \cap \Gamma_0 = \phi \qquad$ and $\qquad \bar{S}^\pm_T \cap S^0_T = \phi$, respectively,

and in the elliptic case

$(H2)_e$ $\quad \inf\limits_{x \in \Omega} c(x) \geq c_0$ with sufficiently large c_0 .

4.3.1. Uniform asymptotic solutions in the case $\Gamma = \Gamma_+$

First, we shall study the elliptic case. The global formal solution

$$u^N(x,\varepsilon) = \sum_{\mu=0}^{N} \varepsilon^\mu u_\mu(x)$$

satisfies

$$L_0 u_0 = f \,, \qquad L_0 u_\mu = - L_1 u_{\mu-1} \,, \qquad \mu = 1,\ldots,N \,.$$

Because of $\partial\Omega = \Gamma_+$ and $(H2)_e$ there are uniquely determined regular solutions u_μ according to Chapter 3. Let $(\varrho,\varphi) = (\varrho, \varphi_1,\ldots,\varphi_{n-1})$ be local coordinates in a neighbourhood of $\partial\Omega$ such that $\varrho(x) = dist(x,\partial\Omega)$ and $0 < \varrho < \varrho_0$ corresponds to a strip parallel to $\partial\Omega$.

L_ε transforms into

$$\tilde{L}_\varepsilon = \varepsilon \sum_{l+|\beta|\le 2m} A_{l\beta}(\rho,\varphi) D_\rho^l D_\varphi^\beta + B_0(\rho,\varphi) D_\rho + \sum_{\mu=1}^{n-1} B_\mu(\rho,\varphi) \frac{\partial}{\partial \varphi_\mu} + C(\rho,\varphi) .$$

Because of $B_0(0,\varphi) < 0$ we set

$$\rho = \varepsilon^{\frac{1}{2m-1}} \zeta = \lambda \zeta$$

and

$$v^M(\zeta,\varphi,\lambda) = \sum_{\mu=0}^{M} \lambda^\mu v_\mu(\zeta,\varphi)$$

for the local corrector. We obtain the equations

$$L v_0 = 0 , \qquad L v_\mu = F_\mu(v_0,\dots,v_{\mu-1}) , \qquad \mu = 1,\dots,M$$

where

$$L v \equiv A_{2m,0}(0,\varphi) \frac{\partial^{2m} v}{\partial \zeta^{2m}} + B_0(0,\varphi) \frac{\partial v}{\partial \zeta} .$$

The F_μ are multilinear differential operators, the coefficients of which are polynomials in ζ. The decomposition of the boundary conditions yields, with $M \ge (2m-1)N + m - 1$,

$$\left.\frac{\partial^s v_{(2m-1)l+s}}{\partial \zeta^s}\right|_{\zeta=0} = - \left.\frac{\partial^s u_l}{\partial \rho^s}\right|_{\rho=0} , \qquad l = 0,1,\dots,N; \quad s = 0,1,\dots,m-1$$

$$\left.\frac{\partial^s v_\mu}{\partial \zeta^s}\right|_{\zeta=0} = 0 , \qquad \mu \ne (2m-1)l + s ; \quad l = 0,\dots,N; \quad s = 0,\dots,m-1 .$$

Because of the uniform ellipticity of L_1 it holds that $(-1)^m A_{2m,0}(0,\varphi) > 0$, therefore

$$A_{2m,0}\, \varkappa^{2m} + B_0\, \varkappa = 0$$

has exactly m roots with negative real part. Consequently, uniquely determined solutions v_μ of boundary layer type do exist. (see A 1.1). Let $\psi \in C^\infty(\bar{\Omega})$ be again a cut-off function and

$$\phi_\varepsilon^N(x) = u^N(x,\varepsilon) + \psi(x)\, v^{(2m-1)(N+1)}(\zeta,\varphi,\lambda) .$$

Then it holds

<u>Theorem 4.9.</u> Under the assumptions (H1), $(H2)_e$ $\phi_\varepsilon^N(x)$ is a uniform asymptotic solution of $(4.7)_\varepsilon$ in the case $\partial\Omega = \Gamma_+$. It holds

$$\| u_\varepsilon - \phi_\varepsilon^N \| \le K \varepsilon^{N+1} . \qquad \blacksquare$$

Proof. The a-priori estimate (Theorem 2.9)

$$\|u_\varepsilon\| \le K\,\varepsilon^{\frac{1+\alpha}{2m-1}}\,\|f\|_\alpha + K\,\varepsilon^{-\frac{n}{4m-2}}\,\|f\|$$

deduced in Chapter 2 is fundamental for the proof. We consider the difference

$$z_\varepsilon = u_\varepsilon - \phi_\varepsilon^{N+\tilde N}$$

for sufficiently large N. According to the construction of $\phi_\varepsilon^{N+\tilde N}$, z_ε satisfies the boundary value problem

$$\begin{aligned} L_\varepsilon z_\varepsilon &= r_\varepsilon && \text{in } \Omega \\ \frac{\partial^s z_\varepsilon}{\partial \nu^s} &= 0 && \text{on } \partial\Omega, \quad s = 0,1,\dots,m-1 \end{aligned}$$

where

$$\|r_\varepsilon\| + \varepsilon^{\frac{\alpha+\frac{n}{2}}{2m}}\,\|r_\varepsilon\|_\alpha \le K\,\varepsilon^{N+\tilde N+1} .$$

Applying the a-priori estimate for sufficiently large $\tilde N$ we obtain

$$\|z_\varepsilon\| \le K\,\varepsilon^{N+1} .$$

Consequently,

$$\|u_\varepsilon - \phi_\varepsilon^N\| \le \|u_\varepsilon - \phi_\varepsilon^{N+\tilde N}\| + \|\phi_\varepsilon^N - \phi_\varepsilon^{N+\tilde N}\| \le K\,\varepsilon^{N+1} \;\blacksquare$$

Remark. We did not use the estimate of Theorem 2.9 completely. Naturally, it is also possible to obtain asymptotic approximations for the derivatives of u_ε. $\blacksquare$

The procedure in the parabolic case is entirely analogous. The global formal solution

$$u^N(x,t,\varepsilon) = \sum_{\mu=0}^{N} \varepsilon^\mu u_\mu(x,t)$$

satisfies

$$\begin{aligned} &\frac{\partial u_0}{\partial t} + L_0 u_0 = f\,, && \frac{\partial u_\mu}{\partial t} + L_0 u_\mu = -L_1 u_{\mu-1}\,, \qquad \mu = 1,\dots,N \\ &u_0\big|_{t=0} = 0\,, && u_\mu\big|_{t=0} = 0 \;. \end{aligned}$$

As above we determine the local corrector, only one F_μ changes. The sharpened compatibility condition ($\tilde{\text{H}}$3) automatically guarantees that $v_\mu\big|_{t=0} = 0$. So, we obtain a result analogous to that of Theorem 4.9 from the a-priori estimate stated for the parabolic case (see Theorem 2.11) $\blacksquare$

4.3.2. Uniform asymptotic solutions in the case $\Gamma = \Gamma_-$

In difference to the previous case we cannot determine the global solution independently of the local corrector.
In the elliptic case let the global solution be

$$u^N(x,\varepsilon) = \sum_{\mu=0}^{N} \lambda^\mu u_\mu(x) , \qquad \lambda^{2m-1} = \varepsilon$$

where the u_μ satisfy

$$L_0 u_\mu = \begin{cases} f & \mu = 0 \\ 0 & \mu = 1,\dots,2m-2 \\ -L_1 u_{\mu-2m+1} , & \mu = 2m-1,\dots,N . \end{cases}$$

Now, additional conditions on Γ_- are proper. We require $u_0|_{\Gamma_-} = 0$, however, the conditions for the other u_μ depend on the local corrector. Because u_0 already satisfies one boundary condition, we set

$$v^M(\zeta,\varphi,\lambda) = \lambda \sum_{\mu=0}^{M-1} \lambda^\mu v_\mu(\zeta,\varphi) \qquad \text{with} \quad M = N+m-1$$

and obtain ordinary differential equations for the v_μ as in 4.3.1. The decomposition of the additional conditions yields

$$u_\mu|_{\Gamma_-} = -v_{\mu-1}|_{\zeta=0} , \qquad \mu = 1,\dots,N$$

and

$$\frac{\partial^s v_\mu}{\partial \zeta^s}\Big|_{\zeta=0} = \begin{cases} -\dfrac{\partial^s}{\partial \nu^s} u_{\mu-s+1} & 0 \le \mu-s+1 \le N \\ 0 , \ \mu-s+1 > N & \text{or} \ \mu-s+1 < 0 , \quad s = 1,\dots,m-1 . \end{cases}$$

Consequently, we determine the u_μ, v_μ in the sequence

$$u_0, v_0, u_1, v_1, \dots, u_N, v_N, v_{N+1}, \dots, v_{N+m-2} .$$

This is uniquely possible because the characteristic equation

$$A_{2m,0}\, \varkappa^{2m} + B_0 \varkappa = 0$$

has exactly m-1 roots with negative real part, here it holds $B_0 > 0$.
Setting

$$\phi_\varepsilon^N(x) = u^N(x,\varepsilon) + \psi(x)\, v^{N+m-1}(\zeta,\varphi,\lambda)$$

we can prove, as above,

<u>Theorem 4.10.</u> Under the assumptions (H1), $(H2)_e$ $\phi_\varepsilon^N(x)$ is a uniform asymptotic solution of $(4.7)_e$ in the case $\partial\Omega = \Gamma_-$. It holds that

$$\|u_\varepsilon - \phi_\varepsilon^N\| \le K \lambda^{N+1} \qquad \text{with } \lambda^{2m-1} = \varepsilon \qquad \blacksquare$$

Remarks. (i) If $\partial\Omega = \Gamma$, it especially holds

$$\| u_\varepsilon - u_0 \| \leq K \varepsilon^{\frac{1}{2m-1}} ,$$

consequently the solution of the reduced problem is a uniform asymptotic solution.

(ii) Again it is possible to give more general estimates for the difference $u_\varepsilon - \phi_\varepsilon^N$. ∎

In the parabolic case the same procedure is connected with difficulties. Namely, the solutions of

$$\frac{\partial u_\mu}{\partial t} + L_0 u_\mu = \begin{cases} f & \mu = 0 \\ 0 & \mu = 1,\dots,2m-2 \\ -L_1 u_{\mu-2m+1} , & \mu = 2m-1,\dots,N \end{cases}$$

$$u_\mu \big|_{t=0} = 0 , \qquad u_\mu \big|_{S_T} = \begin{cases} 0 & \mu = 0 \\ -v_{\mu-1}\big|_{\tau=0} , & \mu = 1,\dots,N \end{cases}$$

are not longer arbitrary regular. Assuming the u_μ to be sufficiently regular (then f must be identically zero in a neighbourhood of $\partial\Omega \times \{0\}$), we obtain Theorem 4.10 also in the parabolic case. For problems of second order we have shown in Section 4.1, using generalized maximum principles, that

$$\| u_\varepsilon - u_0 \| \leq K \varepsilon^{1/2} .$$

Now, for n = 1 we prove that at least $\lim_{\varepsilon \to 0} u_\varepsilon = u_0$ uniformly holds. A proof of this fact which is probably valid for arbitrary n is still open. Therefore, let now n = 1, $\Omega = (0,1)$, $L_0 u = b(x) \frac{\partial u}{\partial x} + cu$ with $b(0) > 0$, $b(1) < 0$ and let c be a constant, for simplicity. Setting $u_\mu = -v_{\mu-1} + u_\mu^*$ $(\mu \geq 1)$, we obtain problems of the form

$$\frac{\partial w}{\partial t} + b(x)\frac{\partial w}{\partial x} + c w = d(x,t)$$
$$w\big|_{t=0} = w\big|_{x=0} = w\big|_{x=1} = 0$$

for u_0 and for the u_μ^*. Let $X(t,\eta,\xi)$ be the solution of

$$\frac{dx}{dt} = b(x) \quad \text{with} \quad x(\eta) = \xi .$$

Assume that D_1 is the domain between $x = X(t,0,0)$ and the positive t-axis, D_2 is the domain bounded by $X(t,0,0)$, $t = 0$ and $X(t,0,1)$ and D_3 is the domain between $X(t,0,1)$ and $x = 1 (t \geq 0)$, then we have

$$w(x,t) = \int_{\delta_1(x,t)}^{t} d(X(\sigma,t,x),\sigma)\, e^{c(\sigma-t)}\, d\sigma \qquad \text{in } D_1$$

$$w(x,t) = \begin{cases} \int_0^t d(X(\sigma,t,x),\sigma)\, e^{c(\sigma-t)}\, d\sigma & \text{in } D_2 \\ \int_{\delta_3(x,t)}^t d(X(\sigma,t,x),\sigma)\, e^{c(\sigma-t)}\, d\sigma & \text{in } D_3 \, . \end{cases}$$

Thereby $\delta_1(\eta,\xi)$ is the t-value for which $X(\delta_1,\eta,\xi) = 0$, analogously, $\delta_3(\eta,\xi)$ is the t-value for which $X(t,\eta,\xi)$ intersects $x = 1$. From this explicit solution formulas we can see in particular the fact already above mentioned that w is arbitrarily regular if $d(x,t)$ is identically zero in neighbourhoods of (0,0) and (1,0).
Now let $\widetilde{\psi}(x,t)$ be a infinitely differentiable function on Q with

$$\widetilde{\psi}(x,t) = \begin{cases} 1 & \text{in } \{0<x<\delta\}\times\{0<t<\delta\} \cup \{0<1-x<\delta\}\times\{0<t<\delta\} \\ 0 & \text{in } \{t>2\delta\} \cup \{2\delta<x<1-2\delta\} \\ 0 \le \widetilde{\psi} \le 1 & \text{otherwise} \, . \end{cases}$$

We decompose f into

$$f = f_1 + f_2$$

with

$$f_1 = (1-\widetilde{\psi})f \, , \qquad f_2 = \widetilde{\psi} f \, ,$$

analogously, let

$$u_\varepsilon = u_\varepsilon^{(1)} + u_\varepsilon^{(2)} \, .$$

Now $u_\varepsilon^{(2)}$ can easily be estimated by an a-priori estimate and for $u_\varepsilon^{(1)}$ we can state an asymptotic expansion corresponding to Theorem 4.10 because f_1 is identically zero in the neighbourhoods of (0,0) and (1,0). Under the assumption ($\widetilde{H}3$) it holds

$$\|u_\varepsilon^{(2)}\| \le K \varepsilon^{1+\frac{\tau^2}{(2m-1)(2m-1-\alpha)} - \frac{\tau+\alpha}{2m}} \delta^{-(\tau-2m+1)} + K \varepsilon^{-\frac{1}{2}\frac{\tau}{(2m-1)(2m-1-\alpha)} - \frac{1}{2\tau}} \delta^{1+[\frac{\tau}{2m}]}$$

according to Theorem 2.1 .
Choosing

$$\delta = \delta(\varepsilon) = \varepsilon^{\frac{1}{2m-1} - \sigma}$$

with

$$\sigma = \frac{1}{3}\left\{ \frac{2}{2m-1} - \frac{m}{(2m-1)(2m-1-\alpha)} - \frac{1}{4m} \right\} \qquad \left(\text{it holds } 0 < \sigma < \frac{1}{2m-1}\right)$$

we now obtain the a-priori estimate for $u_\varepsilon^{(2)}$

$$\|u_\varepsilon^{(2)}\| \le K \varepsilon^{\sigma} \, .$$

Let

$$u_\varepsilon^{(1)}(x,t,\delta) = \sum_{i=0}^{N} \lambda^i u_i(x,t,\delta) + \lambda \sum_{K=0}^{1} \sum_{i=0}^{N+m-2} \lambda^i v_i^{(K)}(x,t,\delta)$$

where the u_i, $v_i^{(0,1)}$ are determined analogously as above, only f must be replaced by f_1. With the help of the explicit representations for the u_i we can successively study the δ-dependence of

$u_0, v_0^{(0,1)}, u_1, \ldots, u_N, v_N^{(0,1)}, \ldots, v_{N+m-2}^{(0,1)}$ and obtain

$$|u_i| \le K\delta^{-i}, \quad \left|\frac{\partial^s u_i}{\partial x^{s_1} \partial t^{s_2}}\right| \le K\delta^{-i-s+1}, \quad s = 1,2,\ldots$$

$$|v_i| \le K\delta^{-i}, \quad \left|\frac{\partial^s v_i}{\partial x^s}\right| \le K\delta^{-i}, \quad \left|\frac{\partial^s v_i}{\partial t^s}\right| \le K\delta^{-i-s}, \quad s = 1,2\ldots .$$

From this it finally follows

$$\| u_\varepsilon^{(1)} - \sum_{i=0}^{N} \lambda^i u_i(x,t,\delta) - \lambda \sum_{i=0}^{N+m-2} \lambda^i v_i^{(0)}(x,t,\delta) - \lambda \sum_{i=0}^{N+m-2} \lambda^i v_i^{(1)}(x,t,\delta) \| \le K\left(\frac{\lambda}{\delta}\right)^{N+1}.$$

Specially because of $\delta = \varepsilon^{\frac{1}{2m-1} - \sigma}$ and $\lambda = \frac{1}{2m-1} > \sigma$, N = 0 yields

$$\| u_\varepsilon^{(1)} - u_0(x,t,\delta) \| \le K\varepsilon^\sigma .$$

As

$$\| u_0(x,t) - u_0(x,t,\delta) \| \le K\varepsilon^\sigma$$

from the triangle inequality we have

<u>Theorem 4.11.</u> Under the assumption ($\tilde{H}3$) the solution u_ε of $(4.7)_P$ satisfies

$$\| u_\varepsilon - u_0 \| \le K\varepsilon^\sigma$$

in the case $S_T = S_T^-$ for n = 1, where

$$\sigma = \frac{1}{3}\left\{ \frac{2}{2m-1} - \frac{m}{(2m-1)(2m-1-\alpha)} - \frac{1}{4m} \right\} . \qquad \blacksquare$$

4.3.3. Uniform asymptotic solutions in the case $\Gamma = \Gamma_0$

The procedure in the case $\partial\Omega = \Gamma_0$ is similar to that in the case $\partial\Omega = \Gamma_+$. Let

$$u^N(x,\varepsilon) = \sum_{\mu=0}^{N} \varepsilon^\mu u_\mu(x)$$

be the global formal solution where the $u_\mu(x)$ satisfy

$$L_0 u_0 = f, \qquad L_0 u_\mu = -L_1 u_{\mu-1}, \qquad \mu = 1,\dots,N .$$

Because of $\partial\Omega = \Gamma_0$ and $(H2)_e$, regular solutions u_μ are uniquely determined. In the neighbourhood of the boundary we now set

$$\rho = \lambda\zeta \quad \text{with} \quad \lambda = \varepsilon^{\frac{1}{2m}} .$$

For the v_μ of the local corrector

$$v^M(\zeta,\varphi,\lambda) = \sum_{\mu=0}^{M} \lambda^\mu v_\mu(\zeta,\varphi)$$

we then obtain equations of the form

$$L v_\mu = A_{2m,0}(0,\varphi)\frac{\partial^{2m} v_\mu}{\partial\zeta^{2m}} + \frac{\partial B_\rho}{\partial\rho}(0,\varphi)\zeta\frac{\partial v_\mu}{\partial\zeta} + \sum_{k=1}^{n-1} B_k(0,\varphi)\frac{\partial v_\mu}{\partial\varphi_k} + C(0,\varphi)v_\mu = F_\mu$$

$$F_0 = 0, \qquad F_\mu = F_\mu(v_0,\dots,v_{\mu-1}), \qquad \mu = 1,\dots,M . \qquad (4.8)_e$$

Now the problem of the case $\partial\Omega = \Gamma_0$ consists in the verification that $(4.8)_e$ has solutions of boundary layer type under boundary conditions of the form

$$\left.\frac{\partial^s v_\mu}{\partial\zeta^s}\right|_{\zeta=0} = G_{s,\mu}(\varphi), \quad s = 0,1,\dots,m-1; \quad \mu = 1,\dots,M .$$

If this can be guaranteed we prove, analogously as in the case $\partial\Omega = \Gamma_+$, that $u^N + \psi v^M$ is a uniform asymptotic solution.
The boundary value problems $(4.8)_e$ will be discussed in A 1.2, A 3.
Now we assume

$(H5)_e$ Let one of the following conditions be fulfilled.

(1) $\frac{\partial B_\rho}{\partial\rho}(0,\varphi) \equiv 0, \qquad B_k(0,\varphi) \equiv 0, \quad k = 1,\dots,n-1$

(2) $\frac{\partial B_\rho}{\partial\rho}(0,\varphi) \equiv 0, \quad B_k(0,\varphi) \equiv 0, \; k = 1,\dots,n-1, \; k \neq \ell, \quad B_\ell(0,\varphi) \not\equiv 0$

$A_{2m,0}(0,\varphi)$ and $C(0,\varphi)$ be independent of φ_ℓ.

Then there exist solutions of boundary layer type and we have

Theorem 4.12$_e$. Under the assumptions (H1), (H2)$_e$, (H5)$_e$,

$$\phi_\varepsilon^N = u^N + \psi(x)\, v^{2m(N+1)}$$

is a uniform asymptotic solution of (4.7)$_e$ in the case $\partial\Omega = \Gamma_0$. It holds that

$$\| u_\varepsilon - \phi_\varepsilon^N \| \leq K \varepsilon^{N+1} \quad \blacksquare$$

Remark. Theorem 4.9 and Theorem 4.12 differ in the following way. We have the estimates

$$\| u_\varepsilon - u_0 - \psi v_0 \| \leq K \varepsilon^{\frac{1}{2m-1}} \qquad \text{if} \quad \partial\Omega = \Gamma_+ ,$$

$$\| u_\varepsilon - u_0 - \psi v_0 \| \leq K \varepsilon^{\frac{1}{2m}} \qquad \text{if} \quad \partial\Omega = \Gamma_0 \quad \blacksquare$$

In the parabolic case let

$$u^N(x,t,\varepsilon) = \sum_{\mu=0}^{N} \varepsilon^\mu u_\mu(x,t)$$

where the u_μ satisfy

$$\frac{\partial u_0}{\partial t} + L_0 u_0 = f ,$$

$$\frac{\partial u_\mu}{\partial t} + L_0 u_\mu = - L_1 u_{\mu-1} , \qquad \mu = 1,\dots,N ,$$

$$u_\mu \big|_{t=0} = 0 .$$

For the coefficients of the expansion of the local corrector we obtain problems of the form

$$\frac{\partial v_\mu}{\partial t} + L v_\mu = F_\mu ,$$

$$v_\mu \big|_{t=0} = 0 , \qquad (4.8)_p$$

$$\frac{\partial^s v_\mu}{\partial \zeta^s}\Big|_{\zeta=0} = G_{s,\mu}(\varphi) .$$

Transformations of the form $v = v^* e^{-\vartheta t}$, $t^* = \int_0^t A_{2m,0}\, d\tau$ guarantee that, without restriction of generality, $C \equiv 0$ and $A_{2m,0}(0,\varphi) \equiv (-1)^m$. Analogous to the elliptic case we assume

(H5)$_p$ Let one of the following conditions be fulfilled.

(1) $\frac{\partial B_0}{\partial \rho}(0,\varphi) \equiv 0 , \quad B_k(0,\varphi) \equiv 0 , \quad k = 1,\dots,n-1 ,$

(2) $\frac{\partial B_0}{\partial \rho}(0,\varphi) \equiv 0 , \quad B_k(0,\varphi) \equiv 0 ,\ k = 1,\dots,n-1,\ k \neq k_0 , \quad B_{k_0}(0,\varphi) \neq 0 .$

Then according to A 2, there exist solutions of boundary layer type of (4.8)$_p$ and we have

Theorem 4.12$_p$. Under the assumptions (H1), ($\tilde{H}3$), (H5)$_p$,

$$\phi_\varepsilon^N = u^N + \psi(x)\, V^{2m(N+1)}$$

is a uniform asymptotic solution of $(4.7)_p$ in the case $\partial\Omega = \Gamma_0$.
It holds that

$$\| u_\varepsilon - \phi_\varepsilon^N \| \leq K\, \varepsilon^{N+1} \qquad \blacksquare$$

4.4. Tangential characteristics

In the Sections 4.2 and 4.3 we have always assumed that Γ_- is closed in Γ. This assumption was needed to guarantee regularity of the formal global solution according to Theorem 3.1. In the present section we shall study neighbourhoods of a point at which a characteristic of L_0 is tangent to the domain Ω from outside. Obviously, Γ_- is not closed in Γ then and (H1) is not fulfilled. We shall restrict our considerations to typical examples of equations of second order.

4.4.1. The elliptic case

We consider the boundary value problem

$$\begin{aligned} L_\varepsilon u_\varepsilon \equiv -\varepsilon\, \Delta u_\varepsilon - y \frac{\partial u_\varepsilon}{\partial x} + \frac{\partial u_\varepsilon}{\partial y} &= f(x,y) \qquad \text{in } \Omega, \\ u_\varepsilon &= 0 \qquad \text{on } \partial\Omega \end{aligned} \tag{4.9}$$

in a bounded domain $\Omega \subset \mathbb{R}^2$. Let $\{(x,y) \in \mathbb{R}^2 \mid x = 0\} \cap \partial\Omega \neq \phi$ and $\Omega \subset \{(x,y) \in \mathbb{R}^2 \mid x > 0\}$.
The characteristics $x + \frac{y^2}{2} =$ const. of the operator of first order are tangent to the boundary $x = 0$ at the point $(0,0)$, therefore from outside. It holds $\Gamma_+ \subset \{(0,y) \in \mathbb{R}^2 \mid y > 0\}$, $\Gamma_- \subset \{(0,y) \in \mathbb{R}^2 \mid y < 0\}$ and $\Gamma_0 \subset \{(0,0)\}$. Consequently, (H1) is not fulfilled. We construct uniform asymptotic solutions of (4.9) in a neighbourhood of the origin, more precisely, in

$$\Omega_\kappa = \{ (x,y) \in \mathbb{R}^2 \mid x + \frac{y^2}{2} < \frac{\kappa^2}{2},\ x > 0 \} .$$

First we begin as usual. The global formal solution

$$u^N(x,y,\varepsilon) = \sum_{\mu=0}^{N} \varepsilon^\mu u_\mu(x,y)$$

satisfies

$$-y \frac{\partial u_0}{\partial x} + \frac{\partial u_0}{\partial y} = f(x,y)$$

$$-y \frac{\partial u_\mu}{\partial x} + \frac{\partial u_\mu}{\partial y} = \Delta u_{\mu-1}, \qquad \mu = 1,\dots,N$$

$$u_\mu = 0 \quad \text{on } \Gamma_-, \qquad \mu = 0,1,\dots,N .$$

The u_μ are uniquely determined, but have singularities at (0,0). For instance in the special case $f \equiv 1$ we have

$$u_0(x,y) = y + \sqrt{2x+y^2} .$$

We make use of following notations

$$\theta = x + \frac{y^2}{2}, \qquad R_m(\theta,y) = \sum_{i=0}^{i_m} a_{im}\,\theta^{\frac{m-i}{2}} y^i \qquad (a_{im} \text{ constants}).$$

Let $Q_m(\theta,y)$ be a $R_m(\theta,y)$ with $Q_m(\theta,-\sqrt{2\theta}) = 0$. Then estimates of the form $|R_m(\theta,y)| \le K\,\theta^{\frac{m}{2}}$ are satisfied. The singularities of $u_\mu(x,y)$ are characterized by

<u>Lemma 4.3.</u> For sufficiently large M it holds that

$$u_\mu(x,y) = \sum_{k=1-3\mu}^{M-3\mu} Q_k^{(\mu)}(\theta,y) + U_M^{(\mu)}(x,y), \qquad \mu = 0,1,\dots,N$$

where

$$\frac{\partial}{\partial y} Q_k^{(0)}(\theta,y) = R_{k-1}^{(0)}(\theta,y), \qquad k = 1,\dots$$

$$\frac{\partial}{\partial y} Q_k^{(\mu)}(\theta,y) = \frac{\partial^2}{\partial\theta^2} Q_{k+3}^{(\mu-1)}(\theta,y) + L^0 Q_{k+1}^{(\mu-1)}(\theta,y), \qquad \mu = 1,\dots$$

$$Q_k^{(\mu)}(\theta,y) \equiv 0 \qquad \text{for } k < 1-3\mu$$

$$\left| \frac{\partial^l U_M^{(\mu)}(x,y)}{\partial x^{l_1}\,\partial y^{l_2}} \right| \le K\,\theta^{\frac{M+1-3\mu-2l_1-l_2}{2}}$$

$$L^0 Q \equiv y^2 \frac{\partial^2 Q}{\partial\theta^2} + 2y \frac{\partial^2 Q}{\partial\theta\,\partial y} + \frac{\partial^2 Q}{\partial y^2} + \frac{\partial Q}{\partial\theta} . \quad \blacksquare$$

<u>Proof.</u> For the proof of the lemma we notice that the $Q_k^{(\mu)}$ are uniquely determined by the given relations if $R_{k-1}^{(0)}$ is known. Expanding $f(x,y)$ into a Taylor-series at (0,0) and substituting $x = \theta - \frac{y^2}{2}$ we obtain

$$f(x,y) = \sum_{\mu=0}^{M-1} R_\mu^{(0)}(\theta,y) + F_M^{(0)}(x,y)$$

with

$$|D^l F_M^{(0)}(x,y)| \le K\,\theta^{\frac{M-2l_1-l_2}{2}}, \qquad l = (l_1,l_2) .$$

The $U_M^{(\mu)}$ satisfy problems of the form

$$\left(-y\frac{\partial}{\partial x} + \frac{\partial}{\partial y}\right) U_M^{(\mu)}(x,y) = F_M^{(\mu)}(x,y)$$

$$U_M^{(\mu)}(0,y) = 0 \qquad \text{for } y < 0 .$$

The representation of the solution $U_M^{(\mu)}$ yields the assertion of the lemma.

With the help of Lemma 4.3 it is possible now to estimate $L_\varepsilon(u_\varepsilon - u^N) = \varepsilon^{N+1}\Delta u_N$ in the domain $\Omega_\kappa \setminus \Omega^1$ with

$$\Omega^1 = \{ (x,y) \in \mathbb{R}^2 \mid 0 \le x \le \tfrac{\varepsilon^{2\alpha}}{2}, \ |y| \le \tfrac{\varepsilon^{\alpha}}{2} \} .$$

Because, for Q_μ, estimates of the form $|Q_\mu(\theta,y)| \le K\theta^{\frac{\mu}{2}}$ are valid, we obtain

$$|L_\varepsilon(u_\varepsilon - u^N)| \le K\,\varepsilon^{(N+1)(1-3\alpha)} \qquad in\ \Omega_\kappa \setminus \Omega^1 \tag{4.10}$$

for $0 < \alpha < \frac{1}{3}$.

In the neighbourhood of Γ_+ we study the local corrector

$$v^N(\zeta,y,\varepsilon) = \sum_{\mu=0}^{N} \varepsilon^\mu v_\mu(\zeta,y) , \qquad \zeta = \frac{x}{\varepsilon}$$

with

$$L v_\mu \equiv \frac{\partial^2 v_\mu}{\partial \zeta^2} + y\frac{\partial v_\mu}{\partial \zeta} = \frac{\partial v_{\mu-1}}{\partial y} - \frac{\partial^2 v_{\mu-2}}{\partial y^2} , \quad \mu = 0,1,\dots$$

$$v_\mu(\zeta,y) \equiv 0 \qquad for\ \ \mu < 0$$

$$v_\mu(0,y) = -u_\mu(0,y) , \qquad y > 0, \quad \mu = 0,1,\dots \quad .$$

There exist unique solutions of classical boundary layer type for $y > 0$. To determine the behaviour of the $v_\mu(\zeta,y)$ in the neighbourhood of $y = 0$ we introduce functions of the following form

$$S_m(\zeta,y) = \exp(-\zeta y) \sum_{i=i_0 \ge 0}^{i_m} a_{im}\, y^{m+i} \zeta^i \quad .$$

Furthermore let $T_m(\zeta,y)$ be a $S_m(\zeta,y)$ with $i_0 > 0$. The behaviour of $v_\mu(\zeta,y)$ is characterized by

<u>Lemma 4.4.</u> For sufficiently large M it holds that

$$v_\mu(\zeta,y) = \sum_{\kappa=1-3\mu}^{M-3\mu} S_\kappa^{(\mu)}(\zeta,y) + V_M^{(\mu)}(\zeta,y) , \qquad \mu = 0,1,\dots$$

where

$$L S_\kappa^{(\mu)} = \frac{\partial}{\partial y} S_{\kappa+3}^{(\mu-1)} - \frac{\partial^2}{\partial y^2} S_{\kappa+4}^{(\mu-2)}$$

$$S_\kappa^{(\mu)}(0,y) = -Q_\kappa^{(\mu)}(\tfrac{y^2}{2},y) , \qquad y > 0$$

$$S_\kappa^{(\mu)}(\zeta,y) \equiv 0 \qquad for\ \ \mu<0 \quad or\ \ \kappa < 1-3\mu$$

$$\left| \frac{\partial^l V_M^{(\mu)}(\zeta,y)}{\partial \zeta^{l_1} \partial y^{l_2}} \right| \le K\, y^{M+1-3\mu+l_1-l_2} \exp(-\gamma\zeta y), \ \zeta \ge 0,\ y<1,\ 0<\gamma<1.$$

■

Proof. The proof is carried out by induction. Because $V_0(\zeta,y) = -u_0(0,y)\exp(-\zeta y)$, from Lemma 4.3 we obtain

$$V_0(\zeta,y) = -\sum_{K=1}^{M} Q_K^{(0)}(\tfrac{y^2}{2}, y)\exp(-\zeta y) - U_M^{(0)}(0,y)\exp(-\zeta y).$$

For $S_K^{(0)}(\zeta,y) = -Q_K^{(0)}(\frac{y^2}{2},y)\exp(-\zeta y)$ the assertion follows in the case $\mu = 0$. Now, let the assertion be fulfilled for $\mu = 0,1,\dots,m$. Because $\frac{\partial}{\partial y} S_{K+3}^{(m)} - \frac{\partial^2}{\partial y^2} S_{K+4}^{(m-1)}$ is a S_{K+2}, there is exactly one $T_K^{(m+1)}$ with

$$L\,T_K^{(m+1)} = \frac{\partial}{\partial y} S_{K+3}^{(m)} - \frac{\partial^2}{\partial y^2} S_{K+4}^{(m-1)}.$$

Setting

$$V_{m+1}(\zeta,y) = \sum_{K=1-3m-3}^{M-3m-3} \left[-Q_K^{(m+1)}(\tfrac{y^2}{2},y)\exp(-\zeta y) + T_K^{(m+1)}(\zeta,y)\right] + V_M^{(m+1)}(\zeta,y)$$

we obtain

$$L V_M^{(m+1)} = \frac{\partial}{\partial y} V_M^{(m)} - \frac{\partial^2}{\partial y^2} V_M^{(m-1)} + S_{M-3m}$$

$$V_M^{(m+1)}(0,y) = -U_M^{(m+1)}(0,y), \qquad y > 0$$

for the determination of $V_M^{(m+1)}(\zeta,y)$. With the help of an integral representation of $V_M^{(m+1)}$ and Lemma 4.3 we conclude the assertion of the lemma for $\mu = m+1$ ∎

Now we use Lemma 4.4 to estimate $|L_\varepsilon V^N|$ in the case $y \geq \frac{\varepsilon^\alpha}{2}$. From $|S_m(\zeta,y)| \leq K y^m \exp(-\gamma\zeta y)$ it immediately follows

$$|L_\varepsilon V^N| \leq K\varepsilon^{N(1-3\alpha)} \qquad \text{for } x \geq 0,\ y \geq \frac{\varepsilon^\alpha}{2}. \qquad (4.11)$$

To construct a local approximation in a neighbourhood of the origin we introduce the local variables $\xi = \frac{x}{\varepsilon^{2/3}}$, $\eta = \frac{y}{\varepsilon^{1/3}}$ generating the significant degeneration (see Chapter 1)

$$\tilde{L} = \frac{\partial}{\partial \eta} - \frac{\partial^2}{\partial \xi^2} - \eta\frac{\partial}{\partial \xi}$$

of L_ε. Analogous to $\theta = x + \frac{y^2}{2}$ let $\sigma = \xi + \frac{\eta^2}{2}$, consequently, $\theta = \varepsilon^{2/3}\sigma$. Expanding $f(x,y)$ in the new variables we obtain

$$\left| f(x,y) - \sum_{j=0}^{N-1} \varepsilon^{j/3} R_j^{(0)}(\sigma,\eta)\right| \leq K\varepsilon^{N/3}\sigma^{N/2}$$

because $R_j^{(0)}(\theta,y) = \varepsilon^{j/3} R_j(\sigma,\eta)$.

Then the local formal solution

$$w^{3N}(\xi,\eta,\varepsilon) = \sum_{\mu=1}^{3N} \varepsilon^{\mu/3} w_\mu(\xi,\eta)$$

satisfies the relations

$$\begin{aligned}
&\tilde{L} w_\mu = R^{(0)}_{\mu-1} + \frac{\partial^2 w_{\mu-2}}{\partial \eta^2} , \qquad \mu = 1, 2, \ldots \\
&w_\mu(0,\eta) = 0 \\
&w_\mu(\xi,\eta) \equiv 0 \quad \text{for } \mu < 0 \quad .
\end{aligned} \tag{4.12}$$

The solutions of the boundary layer problems (4.12) are not uniquely determined. We obtain the required additional conditions by comparing w^{3N} with the global formal solution u^N. With the help of Lemma 4.3 we express u^N in the new variables ξ and η. It holds

$$u^N(x,y,\varepsilon) = \sum_{K=1}^{M} \varepsilon^{\frac{K}{3}} \sum_{\mu=0}^{N} Q^{(\mu)}_{K-3\mu}(\sigma,\eta) + \sum_{\mu=0}^{N} \varepsilon^\mu \, \mathcal{U}^{(\mu)}_M(x,y) \tag{4.13}$$

where, for the second term, the estimates

$$\left| \sum_{\mu=0}^{N} \varepsilon^\mu \, \mathcal{U}^{(\mu)}_M(x,y) \right| \le K \, \varepsilon^{(M+1)\alpha}$$

$$\left| \frac{\partial}{\partial x} \sum_{\mu=0}^{N} \varepsilon^\mu \, \mathcal{U}^{(\mu)}_M(x,y) \right| \le K \, \varepsilon^{(M-1)\alpha} \qquad \text{in } \Omega^2$$

$$\left| \frac{\partial}{\partial y} \sum_{\mu=0}^{N} \varepsilon^\mu \, \mathcal{U}^{(\mu)}_M(x,y) \right| \le K \, \varepsilon^{M\alpha}$$

are fulfilled in the domain $\Omega^2 = \{(x,y) \in \mathbb{R}^2 \,|\, 0 \le x \le \varepsilon^{2\alpha} \, , \; |y| \le \varepsilon^\alpha\}$.

Because the $Q^{(\mu)}_{K-3\mu}$ in (4.13) satisfy estimates of the form $|Q^{(\mu)}_{K-\mu}| \le K \, \sigma^{\frac{K-3\mu}{2}}$ and since the local formal solution w^{3N} must correspond to u^N for $\eta \to -\infty$, a suitable additional condition for (4.12) is

$$|w_\mu(\sigma,\eta)| \le K \, \sigma^{\mu/2} \quad \text{for} \quad \eta \le -1 \, , \; \xi \ge 0 \; . \tag{4.14}$$

Statements about existence, uniqueness and asymptotic behaviour of the solution of (4.12) with (4.14) follow from

<u>Lemma 4.5.</u> There are unique infinitely differentiable solutions of (4.12), (4.14). It holds the asymptotic representation

$$w_\mu(\xi,\eta) = \sum_{K=0}^{M-1} Q^{(K)}_{\mu-3K}(\sigma,\eta) + W^{(\mu)}_M(\xi,\eta)$$

$$\left| \frac{\partial^l W^{(\mu)}_M(\xi,\eta)}{\partial \xi^{l_1} \, \partial \eta^{l_2}} \right| \le K \, \sigma^{-\frac{3M+r}{2}}$$

for $\eta \leq -1$ and sufficiently large M where r only depends on l and μ. The uniqueness of the solution of (4.12), (4.14) can be established with the help of the barrier function $(\sqrt{2\sigma}+\eta)\sigma^{\mu/2}$. By reduction to problems of the form

$$\tilde{L} W_M = \phi , \qquad | D^l \phi(\xi,\eta)| \leq K \sigma^{\frac{-m+l_2+2l_1}{2}}$$

$$W_M\Big|_{\xi=0} = 0$$

it is possible to prove the existence of a solution W_M with

$$| D^l W_M(\xi,\eta)| \leq K \sigma^{\frac{-m+l_2+2l_1+1}{2}}$$

in the domain $\{(\xi,\eta) \mid \xi \geq 0,\ \eta \leq -1\}$. Then W_M satisfies a parabolic equation with initial and boundary conditions in $\{(\xi,\eta) \mid \xi \geq 0,\ \eta > -1\}$
(see ILJIN/GORKOV/LELIKOVA (1975), LELIKOVA (1976))∎

Remark. The asymptotic representation in Lemma 4.5 is also valid for fixed η and $\xi \to \infty$ ∎

Later we also need the asymptotic behaviour of $w_\mu(\xi,\eta)$ for $\eta \to \infty$. For this purpose we expand $u^N(x,y,\varepsilon) + v^N(\zeta,y,\varepsilon)$ in the new variables ξ and η. From Lemma 4.4 it follows

$$v^N(\zeta,y,\varepsilon) = \sum_{k=1}^{M} \varepsilon^{\frac{k}{3}} \sum_{\mu=0}^{N} S^{(\mu)}_{k-3\mu}(\xi,\eta) + \sum_{\mu=0}^{N} \varepsilon^{\mu} V^{(\mu)}_M(\zeta,y) \qquad (4.15)$$

with

$$\Big| \sum_{\mu=0}^{N} \varepsilon^{\mu} V^{(\mu)}_M(\zeta,y)\Big| \leq K \varepsilon^{(M+1)\alpha}$$

$$\Big| \frac{\partial}{\partial x} \sum_{\mu=0}^{N} \varepsilon^{\mu} V^{(\mu)}_M(\zeta,y)\Big| \leq K \varepsilon^{(M+2)\alpha-1} \quad in \ \{(x,y) \mid 0 < y \leq \varepsilon^{\alpha}\}$$

$$\Big| \frac{\partial}{\partial y} \sum_{\mu=0}^{N} \varepsilon^{\mu} V^{(\mu)}_M(\zeta,y)\Big| \leq K \varepsilon^{M\alpha} .$$

Comparing w^{3N} and $u^N + v^N$ from (4.13) and (4.15), we find the representation

$$w_\mu(\xi,\eta) = \sum_{k=0}^{M-1} \left[Q^{(k)}_{\mu-3k}(\sigma,\eta) + S^{(k)}_{\mu-3k}(\xi,\eta)\right] + \widetilde{W}^{(\mu)}_M(\xi,\eta) \quad for \ \eta \geq 1 .$$

Then the $\tilde{W}_M$ satisfy parabolic boundary value problems of the form

$$\tilde{L}\tilde{W}_M = \tilde{\phi}$$
$$\tilde{W}_M(0,\eta) = 0, \qquad \tilde{W}_M(\xi,1) = \hat{\phi}(\xi)$$

with

$$|\tilde{\phi}(\xi,\eta)| \le K\{\sigma^{-\frac{m}{2}} + \eta^{-n}\exp(-\gamma\xi\eta)\}, \qquad |\hat{\phi}(\xi)| \le K(2\xi+1)^{\frac{1-m}{2}}$$

and corresponding estimates for the derivatives. With the help of the barrier function

$$K_1(\sqrt{2\sigma}+\eta)\sigma^{-\frac{m}{2}} + K_2\eta^{-(l+2)}\exp(-\gamma\xi\eta)$$

we can estimate $\tilde{W}_M$ and obtain

Lemma 4.6. For $\eta \ge 1$ and sufficiently large M it holds the asymptotic representation

$$w_\mu(\xi,\eta) = \sum_{k=0}^{M-1}\left[Q^{(k)}_{\mu-3k}(\sigma,\eta) + S^{(k)}_{\mu-3k}(\xi,\eta)\right] + \tilde{W}^{(\mu)}_M(\xi,\eta)$$

$$\left|\frac{\partial^l \tilde{W}^{(\mu)}_M(\xi,\eta)}{\partial\xi^{l_1}\partial\eta^{l_2}}\right| \le K\sigma^{-\frac{3M+r}{2}}$$

where r only depends on μ and l. ■

We now compose a uniform asymptotic solution by means of the three constructed formal solutions. Let $\psi(x) \in C^\infty(\bar{\Omega})$ be a cut-off function with $0 \le \psi(x) \le 1$ and

$$\psi(x) = \begin{cases} 1 & \text{for } x \le \frac{1}{2} \\ 0 & \text{for } x \ge 1, \end{cases}$$

and let α be a fixed real number with $0 < \alpha < \frac{1}{3}$. Setting

$$\phi^N_\varepsilon(x,y) = u^N(x,y,\varepsilon)\left[1 - \psi(\tfrac{x}{\varepsilon^{3\alpha}})\psi(\tfrac{y}{\varepsilon^\alpha})\psi(-\tfrac{y}{\varepsilon^\alpha})\right] + v^N(\zeta,y,\varepsilon)\left[1-\psi(\tfrac{y}{\varepsilon^\alpha})\right] +$$
$$+ w^{3N}(\xi,\eta,\varepsilon)\psi(\tfrac{x}{\varepsilon^{3\alpha}})\psi(\tfrac{y}{\varepsilon^\alpha})\psi(-\tfrac{y}{\varepsilon^\alpha})$$

we obtain

Theorem 4.13$_e$. ϕ^N_ε is a uniform asymptotic solution of (4.9) in $\bar{\Omega}_K$. It holds that

$$\max_{\bar{\Omega}_K}|u_\varepsilon(x,y) - \phi^N_\varepsilon(x,y)| \le K\varepsilon^{\vartheta N}$$

where ϑ is a positive real number independent of N and ε. ■

Proof. Because $v^N(\zeta,y,\varepsilon)$ and their derivatives are exponentially small for $x \geq \varepsilon^{2\alpha}$, $\frac{\varepsilon^\alpha}{2} \leq y \leq \varepsilon^\alpha$ it immediately follows from (4.10) and (4.11) that

$$|L_\varepsilon(u_\varepsilon - \phi_\varepsilon)| \leq K\varepsilon^{(1-3\alpha)N} \qquad in\ \Omega_\kappa \setminus \Omega^2 .$$

According to the construction of w_μ, Lemma 4.5 and Lemma 4.6 the estimate

$$|L_\varepsilon(u_\varepsilon - w^{3N})| \leq K\varepsilon^{\vartheta N} \qquad in\ \Omega^2$$

is valid. Thereby ϑ does only depend on α, without restriction of generality let $\vartheta \leq 1 - 3\alpha$. We now establish that in the remaining domain $\Omega^2 \setminus \Omega^1$ it holds that

$$|L_\varepsilon(u_\varepsilon - \phi_\varepsilon^N)| \leq K\varepsilon^{\vartheta N} .$$

For this we subdivide $\Omega^3 = \Omega^2 \setminus \Omega^1$ into the subdomains

$$\Omega^4 = \Omega^3 \cap \{(x,y) \mid y \geq \varepsilon^{1/3}\}$$
$$\Omega^5 = \Omega^3 \cap \{(x,y) \mid y \leq \varepsilon^{1/3}\}$$

and express ϕ_ε^N in the same way as

$$\phi_\varepsilon^N(x,y) = \begin{cases} w^{3N}(\xi,\eta,\varepsilon) + [\psi(\frac{x}{\varepsilon^{2\alpha}})\psi(\frac{y}{\varepsilon^\alpha}) - 1][w^{3N}(\xi,\eta,\varepsilon) - u^N(x,y,\varepsilon) - v^N(\zeta,y,\varepsilon)] + \\ \quad + \psi(\frac{y}{\varepsilon^\alpha})[\psi(\frac{x}{\varepsilon^{2\alpha}}) - 1]\, v^N(\zeta,y,\varepsilon) \qquad in\ \Omega^4 \\ w^{3N}(\xi,\eta,\varepsilon) + [\psi(\frac{x}{\varepsilon^{2\alpha}})\psi(-\frac{y}{\varepsilon^\alpha}) - 1][w^{3N}(\xi,\eta,\varepsilon) - u^N(x,y,\varepsilon)] \quad in\ \Omega^5 \end{cases}$$

where we assume $2\varepsilon^{1/3} < \varepsilon^\alpha$ without restriction of generality. It holds $L_\varepsilon(uv) = uL_\varepsilon v + vL_\varepsilon u - 2\varepsilon\nabla u\cdot\nabla v$. Therefore we also need estimates of v^N, $w^{3N} - u^N - v^N$ and $w^{3N} - u^N$ in the corresponding domains. v^N is exponentially small for $x \geq \frac{\varepsilon^{2\alpha}}{2}$ and $y \geq \varepsilon^{1/3}$. To estimate $w^{3N} - u^N - v^N$ we suppose $M = 3N$ in (4.13) and (4.15) and M to be sufficiently large in Lemma 4.6. In Ω^5 we make use of Lemma 4.5 for sufficiently large M. Summarizing, we have

$$|L_\varepsilon(u_\varepsilon - \phi_\varepsilon^N)| \leq K\varepsilon^{\vartheta N} \qquad in\ \Omega_\kappa$$

$$(u_\varepsilon - \phi_\varepsilon^N)\big|_{x=0} = 0$$

$$|u_\varepsilon - \phi_\varepsilon^N| \leq K \qquad on\ \partial\Omega_\kappa .$$

The function

$$S(x,y,\varepsilon) = K_1(K_2+y)\varepsilon^{\vartheta N} + K_3 \exp\left(\frac{x+\frac{y^2}{2}-\frac{\delta^2}{2}+(K_2+y)\varepsilon^{1/4}}{\varepsilon^{1/2}}\right)$$

majorizes $u_\varepsilon - \phi_\varepsilon^N$ such that the assertion of the theorem is valid in Ω_{κ_1} with $\kappa_1 < \kappa$ ∎

4.4.2. The parabolic case

In particular, we consider

$$\frac{\partial u_\varepsilon}{\partial t} - \varepsilon \frac{\partial^2 u_\varepsilon}{\partial x^2} - t^{r-1}\frac{\partial u_\varepsilon}{\partial x} = f(x,t)$$
$$u_\varepsilon(0,t) = 0 \tag{4.16}$$

in $Q_\varkappa = \left\{(x,t) \;\middle|\; x + \frac{t^r}{r} < \frac{\varkappa^r}{r},\ x>0\right\}$ with an even number $r > 0$.
The characteristics of the first order operator are tangent to the boundary $x = 0$, $-\varkappa \le t \le +\varkappa$ of $Q_\varkappa$ of the order $r - 1$ at $(0,0)$ from outside. The procedure is analogous to that in the elliptic case. The global formal solution

$$u^N(x,t,\varepsilon) = \sum_{\mu=0}^{N} \varepsilon^\mu u_\mu(x,t)$$

satisfies

$$\frac{\partial u_0}{\partial t} - t^{r-1}\frac{\partial u_0}{\partial x} = f(x,t)$$

$$\frac{\partial u_\mu}{\partial t} - t^{r-1}\frac{\partial u_\mu}{\partial x} = \frac{\partial^2 u_{\mu-1}}{\partial x^2}, \qquad \mu = 1,\dots,N$$

$$u_\mu(0,t) = 0 \qquad \text{for } t<0,\ \mu = 0,1,\dots,N,$$

and the local corrector

$$v^N(\zeta,t,\varepsilon) = \sum_{\mu=0}^{N} \varepsilon^\mu v_\mu(\zeta,t), \qquad \zeta = \frac{x}{\varepsilon}$$

again satisfies

$$L v_\mu \equiv \frac{\partial^2 v_\mu}{\partial \zeta^2} + t^{r-1}\frac{\partial v_\mu}{\partial \zeta} = \frac{\partial v_{\mu-1}}{\partial t}, \qquad \mu = 0,1,\dots,N$$

$$v_\mu(\zeta,t) \equiv 0 \qquad \text{for } \mu<0$$

$$v_\mu(0,t) = -u_\mu(0,t) \quad \text{for } t>0,\quad \mu = 0,1,\dots,N$$

such that unique solutions of classical boundary layer type do exist for $t > 0$. The singularities of u_μ and v_μ at the point (0,0) are characterized by Lemma 4.7 and 4.8. Thereby we introduce the relations

$$\theta = x + \frac{t^r}{r}, \qquad R_m(\theta, t) = \sum_{i=0}^{i_1} a_{im}\, \theta^{\frac{m-i}{r}} t^i$$

$$S_m(\zeta, t) = \exp(-\zeta t^{r-1}) \sum_{i=i_0 \geq 0}^{i_1} \tilde{a}_{im}\, \zeta^i\, t^{m+(r-1)i}$$

analogous to the elliptic case. Let $Q_m(\theta,t)$ be a $R_m(\theta,t)$ with

$$Q_m(\theta, -(r\theta)^{\frac{1}{r}}) = 0 .$$

<u>Lemma 4.7.</u> For sufficiently large M, $s = 2r - 1$ it holds

$$u_\mu(x,t) = \sum_{\kappa=1-s\mu}^{M-s\mu} Q_\kappa^{(\mu)}(\theta, t) + U_M^{(\mu)}(x,t)$$

where

$$\frac{\partial}{\partial t} Q_\kappa^{(0)}(\theta, t) = R_{\kappa-1}^{(0)}(\theta, t), \qquad \kappa = 1, \dots, N$$

$$\frac{\partial}{\partial t} Q_\kappa^{(\mu)}(\theta, t) = \frac{\partial^2}{\partial \theta^2} Q_{\kappa+s}^{(\mu-1)}, \qquad \mu \geq 1$$

$$\left| \frac{\partial^l U_M^{(\mu)}(x,t)}{\partial x^{l_1} \partial t^{l_2}} \right| \leq K\, \theta^{\frac{1}{r}(M+1-s\mu-rl_1-l_2)}$$

■

<u>Lemma 4.8.</u> For sufficiently large M, $s = 2r - 1$ it holds

$$v_\mu(\zeta, t) = \sum_{\kappa=1-s\mu}^{M-s\mu} S_\kappa^{(\mu)}(\zeta, t) + V_M^{(\mu)}(\zeta, t)$$

where

$$L\, S_\kappa^{(\mu)} = \frac{\partial}{\partial t} S_\kappa^{(\mu-1)}$$

$$S_\kappa^{(\mu)}(0, t) = -\, Q_\kappa^{(\mu)}\left(\frac{t^r}{r}, t\right) \qquad \text{for } t > 0$$

$$S_\kappa^{(\mu)}(\zeta, t) \equiv 0 \qquad \text{for } \mu < 0$$

$$\left| \frac{\partial^l V_M^{(\mu)}(\zeta,t)}{\partial \zeta^{l_1} \partial t^{l_2}} \right| \leq K\, t^{M+1-s\mu+(r-1)l_1-l_2} \exp(-\gamma \zeta t^{r-1}), \quad 0 < \gamma < 1 .$$

■

By Lemma 4.7 and 4.8 we obtain the estimates

$$|L_\varepsilon(u_\varepsilon - u^N)| \leq K\,\varepsilon^{(1-\alpha s)(N+1)} \qquad \text{in } Q_\kappa \setminus Q^1$$

$$|L_\varepsilon V^N| \leq K\,\varepsilon^{(1-\alpha s)N} \quad . \quad \text{in } Q_\kappa \cap \{(x,y) \mid t \geq \tfrac{\varepsilon^\alpha}{2}\}$$

if we set $Q^1 = \{(x,t) \mid 0 \leq x \leq \frac{\varepsilon^{r\alpha}}{2},\ |t| \leq \frac{\varepsilon^\alpha}{2}\}$ and $0 < \alpha < \frac{1}{s}$.
For the local solution

$$w^N(\xi,\eta,\varepsilon) = \sum_{\mu=1}^{sN} \varepsilon^{\frac{\mu}{s}} w_\mu(\xi,\eta), \qquad \xi = x\,\varepsilon^{-\frac{r}{2r-1}},\ \eta = t\,\varepsilon^{-\frac{1}{2r-1}}$$

we obtain the problems

$$\tilde{L} w_\mu \equiv \frac{\partial w_\mu}{\partial \eta} - \frac{\partial^2 w_\mu}{\partial \xi^2} - \eta^{r-1}\frac{\partial w_\mu}{\partial \xi} = R^{(0)}_{\mu-1}(\sigma,\eta), \qquad \sigma = \xi + \frac{\eta^r}{r}$$

$$w_\mu(0,\eta) = 0, \qquad \mu = 1,\dots,\kappa N \tag{4.17}$$

$$|w_\mu(\xi,\eta)| \leq K\,\sigma^{\frac{\mu}{r}} \qquad \text{for } \eta \leq -1 \quad .$$

We notice that, in difference to the elliptic case, L° from Lemma 4.3, $\frac{\partial^2}{\partial y^2} S^{(\mu-2)}_{\kappa+4}$ from Lemma 4.4 and $\frac{\partial^2}{\partial \eta^2} w_{\mu-2}$ from (4.12) do not appear here. Therefore it is possible to characterize the asymptotic behaviour of the w_μ more precisely.

<u>Lemma 4.9.</u> There are infinitely differentiable solutions of (4.17) in $\xi \geq 0$. For $\eta \leq -1$ and sufficiently large M it holds

$$w_\mu(\xi,\eta) = \sum_{\kappa=0}^{M-1} Q^{(\kappa)}_{\mu-s\kappa}(\sigma,\eta) + W^{(\mu)}_M(\xi,\eta)$$

$$\left|\frac{\partial^l W^{(\mu)}_M(\xi,\eta)}{\partial \xi^{l_1}\,\partial \eta^{l_2}}\right| \leq K\,\sigma^{\frac{1}{r}(-sM+\mu+l_2+r l_1)} \qquad \blacksquare$$

<u>Lemma 4.10.</u> For $\eta \geq 1$ and sufficiently large M it holds

$$w_\mu(\xi,\eta) = \sum_{\kappa=0}^{M-1}\left[Q^{(\kappa)}_{\mu-s\kappa}(\sigma,\eta) + S^{(\kappa)}_{\mu-s\kappa}(\sigma,\eta)\right] + \tilde{W}^{(\mu)}_M(\xi,\eta)$$

$$\left|\frac{\partial^l \tilde{W}^{(\mu)}_M(\xi,\eta)}{\partial \xi^{l_1}\,\partial \eta^{l_2}}\right| \leq K\,\sigma^{\frac{1}{r}(-sM+\mu+l_2+r l_1)} \qquad \blacksquare$$

We now compose a uniform asymptotic solution from the three formal solutions. Let the cut-off function be explained as in the elliptic case and let

$$\phi_\varepsilon^N(x,t) = u^N(x,y,\varepsilon)\left[1-\psi\left(\frac{x}{\varepsilon^{r\alpha}}\right)\psi\left(\frac{t}{\varepsilon^{\alpha}}\right)\psi\left(-\frac{t}{\varepsilon^{\alpha}}\right)\right] + v^N(\zeta,t,\varepsilon)\left[1-\psi\left(\frac{t}{\varepsilon^{\alpha}}\right)\right] +$$

$$+ w^N(\xi,\eta,\varepsilon)\,\psi\left(\frac{x}{\varepsilon^{r\alpha}}\right)\psi\left(\frac{t}{\varepsilon^{\alpha}}\right)\psi\left(-\frac{t}{\varepsilon^{\alpha}}\right) .$$

As in the proof of Theorem 2.13$_e$ we can prove

$$\left|\frac{\partial(u_\varepsilon-\phi_\varepsilon)}{\partial t} + L_\varepsilon(u_\varepsilon-\phi_\varepsilon)\right| \le K\varepsilon^{\vartheta N} \qquad in\ Q_\kappa$$

with $\vartheta = \min(\alpha s,\ 1-\alpha s)$. The best possible choice of α is $\alpha = \frac{1}{2s}$, and we have

Theorem 4.13$_p$. $\phi_\varepsilon^N(x,t)$ is a uniform asymptotic solution of (4.16) in $\bar{Q}_\kappa$. For $\alpha = \frac{1}{2s}$ it holds that

$$\max_{\bar{Q}_\kappa}\left|u_\varepsilon(x,t) - \phi_\varepsilon^N(x,t)\right| \le K\varepsilon^{\frac{N}{2}} \qquad ■$$

4.5. Turning points

We consider the elliptic boundary value problem

$$L_\varepsilon u_\varepsilon \equiv \varepsilon \sum_{i,j=1}^{n} a_{ij} \frac{\partial^2 u_\varepsilon}{\partial x_i \partial x_j} + \sum_{i=1}^{n} b_i \frac{\partial u_\varepsilon}{\partial x_i} + c\, u_\varepsilon = f \quad \text{in } \Omega \qquad (4.18)^*$$
$$u_\varepsilon = 0 \quad \text{on } \partial\Omega$$

and the eigenvalue problem

$$\varepsilon \sum_{i,j=1}^{n} a_{ij} \frac{\partial^2 u_\varepsilon}{\partial x_i \partial x_j} + \sum_{i=1}^{n} b_i \frac{\partial u_\varepsilon}{\partial x_i} = \tau_\varepsilon u_\varepsilon \qquad \text{in } \Omega$$
$$u_\varepsilon = 0 \qquad \text{on } \partial\Omega \,. \qquad (4.19)$$

For $c > 0$ there exists a uniquely determined solution of $(4.18)^*$ with

$$\|u_\varepsilon\| \le K^* \|f\| \,.$$

Consequently, (4.19) has a smallest <u>positive eigenvalue</u> $\hat{\tau}_\varepsilon$ such that for all the other eigenvalues τ_ε it holds that

$$|\operatorname{Re} \tau_\varepsilon| \geq \hat{\tau}_\varepsilon \quad \text{and} \quad |\tau_\varepsilon| > \hat{\tau}_\varepsilon \quad .$$

If all characteristics of L_0 leave the domain $\bar{\Omega}$ in finite time (that means, in particular, that there are no turning points, i. e. b(x) does not vanish in $\bar{\Omega}$), then for any c there exists a uniquely determined solution with

$$\|u_\varepsilon\| \le K^* \|f\|$$

for sufficiently small ε , since, according to Lemma 3.2, there exists a $\eta(x)$ with $b \cdot \nabla\eta > 0$ in $\bar{\Omega}$ and then it holds $c > 0$ without restriction of generality.
Consequently, we have

$$\hat{\tau}_\varepsilon > \frac{K}{\varepsilon}$$

in this case.
If, without restriction of generality, $x = 0 \in \Omega$ is a turning point and $|b(x)| \le K^* |x|^\gamma$ holds in the neighbourhood of $x = 0$, then the estimate

$$\hat{\tau}_\varepsilon \le K \varepsilon^{\frac{2(\gamma-1)}{\gamma+1}}$$

is valid. This means, especially, that

$$\lim_{\varepsilon \to 0} \hat{\tau}_\varepsilon = 0$$

if $\gamma > 1$ and therefore at least one non-simple turning point does exist.
For $b \cdot \nu > 0$ on Γ $\hat{\tau}_\varepsilon$ exponentially converges to zero. In the case of the example

$$\varepsilon \sum_{i,j=1}^{n} a_{ij} \frac{\partial^2 u_\varepsilon}{\partial x_i \partial x_j} + (x_1 \frac{\partial u_\varepsilon}{\partial x_1} + b_2 x_2 \frac{\partial u_\varepsilon}{\partial x_2}) = \tau_\varepsilon u_\varepsilon , \quad \Omega = \{x \mid |x| < 1\}$$

all eigenvalues can be determined, and we obtain

$$|\tau_\varepsilon - (m_1 + b_2 m_2)| \le K \varepsilon^{1/2}, \qquad m_1, m_2 = 0, 1, 2, \ldots \quad .$$

Actually, the solution of the inhomogenous problem

$$-\varepsilon \Delta u_\varepsilon + x_1 \frac{\partial u_\varepsilon}{\partial x_1} + x_2 \frac{\partial u_\varepsilon}{\partial x_2} = 1 , \qquad u_\varepsilon \big|_{|x|=1} = 0 ,$$

is not uniformly bounded, it even holds that

$$|u_\varepsilon| \ge \gamma_1 \exp(\frac{\gamma_2}{\varepsilon}) .$$

If a $\phi \in C^2$ exists such that $\{\phi_{x_i x_j}(0)\}$ is positive definit and

$$b \cdot \nabla \phi \le -K |x|^{1+\gamma} , \quad K > 0 ,$$

then it holds

$$\hat{\tau}_\varepsilon \ge \tilde{K} \varepsilon^{2 \frac{\gamma - 1}{\gamma + 1}} .$$

In the case $\gamma < 1$ we have then $\lim\limits_{\varepsilon \to 0} \hat{\tau}_\varepsilon = \infty$.

For the example

$$\varepsilon \sum_{i,j=1}^{n} a_{ij} \frac{\partial^2 u_\varepsilon}{\partial x_i \partial x_j} - (x_1 \frac{\partial u_\varepsilon}{\partial x_1} + b_2 x_2 \frac{\partial u_\varepsilon}{\partial x_2}) = \tau_\varepsilon u_\varepsilon , \quad \Omega = \{x \mid |x| < 1\}$$

it holds that $\phi = b_2 x_1^2 + x_2^2$, $\gamma = 1$ and

$$\lim_{\varepsilon \to 0} \hat{\tau}_\varepsilon = 1 + b_2 \quad ;$$

for the eigenvalues we obtain that

$$|\tau_\varepsilon - (m_1 + b_2 m_2)| \le K \varepsilon^{1/2}, \qquad m_1, m_2 = 1, 2, \ldots \quad .$$

Whereas many results of section 4.2 (there held $c(x) \ge c_0 > 0$) can be extended to the case $c(x) > 0$ and even to $c(x) > - \lim\limits_{\varepsilon \to 0} \hat{\tau}_\varepsilon$ (but it frequently holds $\lim\limits_{\varepsilon \to 0} \hat{\tau}_\varepsilon = 0$), for $c \equiv 0$ we are confronted with new problems on condition that turning points do exist.
In this chapter we shall study some of these problems.

4.5.1. Isolated turning points

4.5.1.1. An instable singular point in the case $\Gamma = \Gamma_+$

We consider the elliptic boundary value problem

$$L_\varepsilon u_\varepsilon \equiv \varepsilon \sum_{i,j=1}^{n} a_{ij} \frac{\partial^2 u_\varepsilon}{\partial x_i \partial x_j} + \sum_{i=1}^{n} b_i \frac{\partial u_\varepsilon}{\partial x_i} = 0, \qquad u_\varepsilon\big|_\Gamma = g . \tag{4.18}$$

(4.18) has a uniquely determined bounded solution. Let $\Gamma = \Gamma_+$. Furthermore we restrict the behaviour of the characteristics by

(H6) The manifold of the ω-limit sets in Ω consists of the inner point $x = 0$, in a neighbourhood of which it holds that

$$\sum b_i x_i \geq \beta^* |x|^2, \qquad \beta^* > 0 .$$

The formal global problem has the regular, not uniquely determined solution u_0 = const. Since the characteristics start from $x = 0$ we expect that $\lim\limits_{\varepsilon\to 0} u_\varepsilon = \lim\limits_{\varepsilon\to 0} u_\varepsilon(0)$. Naturally, near to the boundary this cannot be uniformly valid. Therefore, we shall again introduce local coordinates there. Let $v_0 + \varepsilon v_1$ be the local corrector obtained by solving the corresponding ordinary boundary layer equations with constant coefficients.

<u>Theorem 4.14.</u> It holds

$$\| u_\varepsilon(x) - u_\varepsilon(0) - \psi v_0 \| \leq K \begin{cases} \varepsilon^{\sigma}, & \sigma < 1,\ n = 2 \\ \varepsilon^{\frac{1}{2(n-1)}}, & n \geq 3 \end{cases}$$

in $\bar{\Omega}$. ∎

<u>Proof.</u> Let $|x| < 2\sqrt{\varepsilon}$ and $y_i = \dfrac{x_i}{\sqrt{\varepsilon}}$. In $|y| < 2$ we have an elliptic differential equation with uniformly bounded coefficients. According to a theorem of MORREY (cf. MIRANDA 1970) then it holds that in any closed domain inside of $|y| < 2$, therefore in particular in $|y| \leq 1$,

$$\left| \frac{\partial u_\varepsilon}{\partial y_i} \right| \leq K .$$

This implies

$$\left| \frac{\partial u_\varepsilon}{\partial x_i} \right| \leq \frac{K}{\sqrt{\varepsilon}} \quad in\ |x| \leq \sqrt{\varepsilon} .$$

According to the mean value theorem it follows

$$| u_\varepsilon(x) - u_\varepsilon(0) | \leq \frac{K}{\sqrt{\varepsilon}}\, \varepsilon^{\alpha} \sqrt{\varepsilon} \leq K \varepsilon^{\alpha} \qquad in\ |x| \leq \varepsilon^{\alpha}\sqrt{\varepsilon} .$$

Naturally, we also have

$$|u_\varepsilon(x) - u_\varepsilon(0) - \psi V_0| \leq K\varepsilon^\alpha \qquad in\ |x| \leq \varepsilon^\alpha\sqrt{\varepsilon}\ .$$

In the remaining domain we estimate $u_\varepsilon(x) - u_\varepsilon(0) - \psi v_0$ by means of a barrier function. First of all let $n \geq 3$ and

$$S(x) = K\varepsilon^\alpha + K\varepsilon^\beta H^*(x)\ ,$$

where

$$H^*(x) = 1 - \tfrac{1}{2} H(x)\left(\varepsilon^\alpha\sqrt{\varepsilon}\right)^{n-2}$$

and H(x) denotes the fundamental solution of $\sum a_{ij}\frac{\partial^2 u}{\partial x_i \partial x_j} = 0$ with the singularity $\frac{1}{|x|^{n-2}}$ at $x = 0$. Estimates for H(x) (cf. MIRANDA (1970)) yield

$$S \geq K\varepsilon^\alpha \qquad for\ |x| = \varepsilon^\alpha\sqrt{\varepsilon}$$

$$S \geq 0 \qquad for\ x \in \partial\Omega$$

$$L_\varepsilon(S) \geq K \begin{cases} \varepsilon^\beta & |x| \geq x_0 \\ \varepsilon^{\beta+1-2\alpha} & \varepsilon^{-\alpha}\sqrt{\varepsilon} \leq |x| \leq x_0 \\ \varepsilon^{\beta+2\alpha(n-2)} & \varepsilon^\alpha\sqrt{\varepsilon} \leq |x| \leq \varepsilon^{-\alpha}\sqrt{\varepsilon} \end{cases}$$

after some calculations. If we choose $\alpha = \frac{1}{2(n-1)}$, $\beta = \frac{1}{n-1}$, then s is a barrier function for $u_\varepsilon(x) - u_\varepsilon(0) - \psi(v_0 + \varepsilon v_1)$ and the assertion for $n \geq 3$ follows from $H^*(x) < 2$.
In the case of $n = 2$ we choose

$$H^*(x) = 1 - \frac{\alpha + 1/2}{\ln\varepsilon} H(x)$$

where H(x) denotes the fundamental solution of $\sum a_{ij}\frac{\partial^2 u}{\partial x_i \partial x_j} = 0$ with the singularity $\ln\frac{1}{|x|}$ at $|x| = 0$. ∎

The determination of $\lim_{\varepsilon\to 0} u_\varepsilon(0)$ is a difficult problem.
For simplicity, we assume

(H7) Let $L_1 = -\Delta$. Further, there exists a regular function P(x) with

$$b(x) = -\nabla\cdot P(x)\ .$$

Then we can write the differential equation in the form

$$\varepsilon\,\nabla\cdot\left(e^{\frac{P}{\varepsilon}}\,\nabla u_\varepsilon\right) = 0$$

and integrating over Ω and then integrating by parts we obtain

$$\int_\Gamma e^{\frac{P}{\varepsilon}}\frac{\partial u_\varepsilon}{\partial\nu}\,d\Gamma = 0$$

(now let ν be the unit normal). In the boundary strip with $\psi \equiv 0$ (for simplicity again $0 < \rho < \rho_0$) let

$$u_\varepsilon = u_\varepsilon(0) + v_0 + \varepsilon v_1 + z_\varepsilon \quad .$$

$\frac{\partial v_0}{\partial \nu}$ and $\frac{\partial v_1}{\partial \nu}$ can be expressed explicitly since v_0 and v_1 are solutions of ordinary differential equations. We obtain

$$\frac{\partial v_0}{\partial \nu} = -(g - u_\varepsilon(0))(b \cdot \nu)\varepsilon^{-1}$$

and

$$\varepsilon \left| \frac{\partial v_1}{\partial \nu} \right| \leq K \ .$$

From this it follows

$$\lim_{\varepsilon \to 0} \int_\Gamma e^{\frac{\rho}{\varepsilon}} (g - u_\varepsilon(0))(b \cdot \nu)\, d\Gamma = 0$$

on condition that

$$\lim_{\varepsilon \to 0} \varepsilon \left| \frac{\partial z_\varepsilon}{\partial \nu} \right| = 0 \ .$$

Corresponding to the construction and to Theorem 4.14 z_ε satisfies

$$|L_\varepsilon z_\varepsilon| \leq K\varepsilon$$

$$z_\varepsilon\Big|_\Gamma = 0 \ , \qquad |z_\varepsilon| \leq K\varepsilon^{\bar{\varepsilon}} \ .$$

Let Ω_0 be the domain $0 < \rho < \rho_0$ and $\Gamma_1 = \partial\Omega_0 \setminus \Gamma$. Suppose that U satisfies

$$L_0 U = K_1 \quad \text{in } \Omega_0 \, , \qquad U\Big|_\Gamma = 0$$

and that $\tilde{v}_0 + \varepsilon\, \tilde{v}_1$ is the local corrector in the neighbourhood of Γ such that

$$\tilde{v}_0 + \varepsilon \tilde{v}_1\Big|_\Gamma = 0 \qquad \text{and} \qquad |L_\varepsilon(\tilde{v}_0 + \varepsilon \tilde{v}_1)| \leq K\varepsilon \ .$$

Then

$$\tilde{z}_\varepsilon = \varepsilon U + K_2(\varepsilon)\left(1 - (\tilde{v}_0 + \varepsilon \tilde{v}_1)\right)$$

satisfies, first of all,

$$\tilde{z}_\varepsilon\Big|_\Gamma = 0 \quad ,$$

further

$$L_\varepsilon \tilde{z}_\varepsilon = K_1\varepsilon - K_2(\varepsilon) L_\varepsilon(\tilde{v}_0 + \varepsilon \tilde{v}_1)$$

and

$$\tilde{z}_\varepsilon\big|_{\Gamma_1} = \varepsilon\, U\big|_{\Gamma_1} + K_2(\varepsilon) - (\tilde{v}_0 + \varepsilon \tilde{v}_1)\big|_{\Gamma_1} .$$

With a sufficient large constant K_1 $L_\varepsilon \tilde{z}_\varepsilon$ majorises the term $|L_\varepsilon z_\varepsilon|$, and by setting

$$K_2(\varepsilon) = K_3 \varepsilon + K_4 \varepsilon^{\tilde{\sigma}}$$

$|z_\varepsilon|_{\Gamma_1}$ is majorized. Furthermore, we have

$$z_\varepsilon \pm \tilde{z}_\varepsilon \geq 0 \qquad in\ \Omega_0$$

$$z_\varepsilon \pm \tilde{z}_\varepsilon\big|_\Gamma = 0 .$$

This implies

$$\frac{\partial}{\partial \nu}(z_\varepsilon \pm \tilde{z}_\varepsilon) \leq 0$$

and consequently

$$\left|\frac{\partial z_\varepsilon}{\partial \nu}\right| \leq \left|\frac{\partial \tilde{z}_\varepsilon}{\partial \nu}\right| \leq K_2(\varepsilon)\,\frac{K_5}{\varepsilon} \quad ;$$

actually it holds that

$$\lim_{\varepsilon \to 0} \varepsilon \left|\frac{\partial z_\varepsilon}{\partial \nu}\right| = 0 .$$

<u>Theorem 4.15.</u> Under the assumptions (H6), (H7) it holds

$$\lim_{\varepsilon \to 0} u_\varepsilon(0) = \lim_{\varepsilon \to 0} \frac{\int_\Gamma e^{\frac{P}{\varepsilon}} g(b\cdot\nu)\, d\Gamma}{\int_\Gamma e^{\frac{P}{\varepsilon}} (b\cdot\nu)\, d\Gamma} . \qquad \blacksquare$$

Let $n = 2$ and

$$b(x) = (b_1 x_1 , b_2 x_2) .$$

Then it holds

$$P(x) = -\tfrac{1}{2}(b_1 x_1^2 + b_2 x_2^2)$$

and P has a maximum at those points at which the largest possible ellipse in Ω touches the boundary $\partial\Omega$. At those points we obtain contributions to the above limit.

If, specially, $b_1 = b_2 = 1$ and if Ω is the unit circle, then we have touching at all boundary points and it holds that

$$\lim_{\varepsilon \to 0} u_\varepsilon(0) = \frac{1}{2\pi} \int_0^{2\pi} g(\varphi)\, d\varphi .$$

4.5.1.2. An instable singular point in the case $\Gamma = \Gamma_0$

Now we consider the case in which, in the interior of the domain, the characteristics behave as in the preceding section but do not intersect the boundary.

Near to the boundary let, as in 4.2.1,

$$L_\varepsilon \equiv \varepsilon \sum_{l+|\beta|\leq 2} A_{l\beta} D_\varrho^l D_\varphi^\beta + B_0(\varrho,\varphi) D_\varrho + \sum_{k=1}^{n-1} B_k(\varrho,\varphi) D_{\varphi_k} \quad .$$

Now we assume

(H8) $B_k(0,\varphi) \equiv 0$ for all k and

$B_0(\varrho,\varphi) \equiv \varrho^l\, B_0^*(\varrho,\varphi)$ with $B_0^* < 0$ for $0 < \varrho < \varrho_0$,

where l is a positive integer number.

If $\Omega = \{x \mid |x| < 1\}$ and $r^2 = \sum_{i=1}^{n} x_i^2$, then, for instance,

$$L_0 \equiv (1-r^2)^l \sum_{i=1}^{n} x_i \frac{\partial}{\partial x_i}$$

satisfies both (H6) and (H8), in particular it holds that

$$B_0(\varrho,\varphi) = -\varrho^l(1-\varrho)(2-\varrho)^l \, .$$

Let v be a local corrector in the neighbourhood of Γ. With $\zeta = \varrho\,\varepsilon^{-\frac{1}{1+l}}$ v satisfies

$$-A\frac{\partial^2 v}{\partial\zeta^2} + B\zeta^l \frac{\partial v}{\partial \zeta} = 0$$
$$v = g - u_\varepsilon(0) \qquad \text{for } \zeta = 0 \quad ,$$

where $A = A_{2,0}(0,\varphi)$, $B = B_0^*(0,\varphi)$.
According to A 1 there exists a uniquely determined solution of classical boundary layer type.

<u>Theorem 4.16.</u> Under the assumptions (H6) and (H8) it holds that

$$\| u_\varepsilon - u_\varepsilon(0) - \psi v \| \leq K\varepsilon^\sigma \quad ,$$

where σ is a positive constant. ∎

<u>Proof.</u> In correspondence to the construction we have

$$| L_\varepsilon(u_\varepsilon - u_\varepsilon(0) - \psi v) | \leq K\varepsilon^{\frac{l}{l+1}} \qquad \text{in } \Omega$$

and

$$u_\varepsilon - u_\varepsilon(0) - \psi v \Big|_{\varrho=0} = 0 \; .$$

In the domain $0 < \varrho < \varrho_0$ we estimate $u_\varepsilon - u_\varepsilon(0) - \psi v$ by means of a special barrier function. For this purpose first of all we need an estimate for $\varrho = \varrho_0$. Because of the assumptions (H6) and (H8), in the domain

$$\Omega_{\frac{\varrho_0}{2}} = \left\{ x \in \Omega \,\middle|\, dist(x, \partial\Omega) > \frac{\varrho_0}{2} \right\}$$

it is possible to use the same procedure as in 4.5.1.1. with a cut-off function ψ^* we obtain

$$\| u_\varepsilon(x) - u_\varepsilon(0) - \psi^* v_0 \| \leq K \varepsilon^{\tilde{\sigma}}$$

in $\Omega_{\frac{\varrho_0}{2}}$, according to Theorem 4.14. From this it immediately follows that

$$|u_\varepsilon(x) - u_\varepsilon(0)| \leq K \varepsilon^{\tilde{\sigma}} \qquad for \ \varrho = \varrho_0$$

and also

$$|u_\varepsilon(x) - u_\varepsilon(0) - \psi v| \leq K \varepsilon^{\tilde{\sigma}} \qquad for \ \varrho = \varrho_0 \ ,$$

since v_0 is a classical boundary layer function.
Now we estimate $u_\varepsilon - u_\varepsilon(0) - \psi v$ in the remaining domain with $0 < \varrho < \varrho_0$ by means of an appropriate barrier function according to the remark following Theorem 2.10.
Let l be even and $l = 2m$.
Setting (cf. Theorem 2.10, case (IV))

$$S = K_1 \varepsilon^{\tilde{\sigma}} - K_2 \varepsilon^{\frac{1}{2m+1}} v_\varepsilon(\varrho)$$

with

$$v_\varepsilon(\varrho) = \int_{-\infty}^{\frac{\varrho}{\varepsilon^{\frac{1}{2m+1}}}} \int_{-\infty}^{q} exp\left(-\frac{B}{A}(p^{2m+1} - q^{2m+1})\right) dp\, dq$$

we obtain

$$|u_\varepsilon - u_\varepsilon(0) - \psi v| \leq S \leq K_1 \varepsilon^{\tilde{\sigma}} + K_2 \varepsilon^{\frac{1}{l+1}} \ .$$

If l is odd and $l = 2m - 1$, we use $S = K_1 \varepsilon^{\tilde{\sigma}} - K_2 \varepsilon^{\frac{1}{2m}} w_\varepsilon(\varrho)$ (Theorem 2.10, case (iii))

$$w_\varepsilon(\varrho) = \int_0^{\frac{\varrho}{\varepsilon^{\frac{1}{2m}}}} \int_0^{q} exp\left(-\frac{B}{A}(p^{2m} - q^{2m})\right) dp\, dq$$

and obtain

$$|u_\varepsilon - u_\varepsilon(0) - \psi v| \leq S \leq K_1 \varepsilon^{\tilde{\sigma}} + K_2 \varepsilon^{\frac{1}{l+1}} \qquad for \ l > 1$$

and

$$|u_\varepsilon - u_\varepsilon(0) - \psi v| \leq S \leq K_1 \varepsilon^{\tilde{\sigma}} + K_2 \varepsilon^{\frac{1}{2}} |ln \, \varepsilon| \qquad for \ l = 1 \ ,$$

respectively.

Therefore, for σ in Theorem 4.16, we obtain

$$\sigma = \min\left(\frac{1}{l+1}, \frac{1}{2(n-1)}\right) \qquad \text{for } n \geq 3, l > 1$$

$$\sigma = \frac{1}{2(n-1)} \qquad \text{for } n = 3, l = 1$$

$$\sigma = \frac{1}{l+1} \qquad \text{for } n = 2, l > 1$$

$$\sigma = \frac{1}{2} - \gamma \quad (\gamma > 0, \text{arbitrary}) \text{ for } n = 2, l = 1 \quad . \qquad \blacksquare$$

To determine $\lim\limits_{\varepsilon \to 0} u_\varepsilon(0)$ we assume again

(H7) Let $L_1 = -\Delta$. Further, there exists a regular $P(x)$ with

$$b(x) = -\nabla P(x) \quad .$$

Then it holds

<u>Theorem 4.17.</u> Under the assumptions (H6), (H7) and (H8) we obtain $\lim\limits_{\varepsilon \to 0} u_\varepsilon(0)$ from

$$\lim_{\varepsilon \to 0} u_\varepsilon(0) = \frac{\int_\Gamma g\, B^* d\Gamma}{\int_\Gamma B^* d\Gamma} , \qquad \text{where } B^* = \left(-B_0^*(0,\varphi)\right)^{\frac{1}{l+1}} \quad . \qquad \blacksquare$$

<u>Proof.</u> Under assumption (H7) we can express the differential equation in the form

$$\varepsilon \nabla \cdot \left(e^{\frac{P}{\varepsilon}} \nabla u_\varepsilon\right) = 0$$

and integrating by parts we obtain

$$\int_\Gamma \nu \cdot \nabla u_\varepsilon \, d\Gamma = \int_\Gamma \frac{\partial u_\varepsilon}{\partial \nu} d\Gamma = 0 \quad ,$$

since P vanishes on Γ (now let ν be the unit normal). From

$$v = (g - u_\varepsilon(0)) \frac{\int_\xi^\infty \exp\left(\frac{B}{l+1} t^{l+1}\right) dt}{\left(-\frac{B}{l+1}\right)^{-\frac{1}{l+1}} \frac{1}{l+1} \Gamma\left(\frac{1}{l+1}\right)}$$

it follows that

$$\left.\frac{\partial v}{\partial \varrho}\right|_{\varrho=0} = -(g - u_\varepsilon(0))\, \varepsilon^{-\frac{1}{l+1}} B^* d(l)$$

with

$$B^* = (-B)^{\frac{1}{l+1}} , \qquad d(l) = (l+1)\left(\frac{1}{l+1}\right)^{\frac{1}{l+1}} \frac{1}{\Gamma\left(\frac{1}{l+1}\right)} \quad .$$

Now let

$$u_\varepsilon = u_\varepsilon(0) + v + z_\varepsilon$$

(in the boundary strip with $\psi \equiv 1$, for instance, again in $0 < \rho < \rho_0$.
Then we immediately obtain the assertion provided that

$$\lim_{\varepsilon \to 0} \varepsilon^{\frac{1}{l+1}} \left| \frac{\partial z_\varepsilon}{\partial \nu} \right| = 0 \quad .$$

According to the construction and Theorem 4.16, z_ε satisfies the estimates

$$|L_\varepsilon z_\varepsilon| \le K \varepsilon^{\frac{l}{l+1}}$$

$$|z_\varepsilon| \le K \varepsilon^{\sigma}, \qquad z_\varepsilon\big|_\Gamma = 0 \, .$$

In the domain $0 < \rho < \rho_0$, we construct an appropriate barrier function such that we can conclude the required estimate for the normal derivative of z_ε .

First, let l be odd and $l = 2m - 1$ as well as $l > 1$. Let $\tilde{z}_\varepsilon$ be defined by

$$\tilde{z}_\varepsilon = -K_1 \varepsilon^{\frac{1}{2m}} w_\varepsilon(\rho) + K_2(\varepsilon)(1 - \tilde{v})$$

where $\tilde{v}$ is a boundary layer function as v with $\tilde{v}\big|_\Gamma \equiv 1$. Obviously, it holds

$$\tilde{z}_\varepsilon\big|_\Gamma = 0 \, .$$

Further, we have

$$L_\varepsilon \tilde{z}_\varepsilon = K_1 \varepsilon^{\frac{2m-1}{2m}} - K_2(\varepsilon) L_\varepsilon(\tilde{v}) \quad \text{with } |L_\varepsilon \tilde{v}| \le K_3 \varepsilon^{\frac{2m-1}{2m}}$$

and

$$\tilde{z}_\varepsilon\big|_{\rho=\rho_0} = -K_1 \varepsilon^{\frac{1}{2m}} w_\varepsilon\big|_{\rho=\rho_0} + K_2(\varepsilon)(1 - \tilde{v}\big|_{\rho=\rho_0}) \, .$$

Choosing

$$K_2(\varepsilon) = K_4 \varepsilon^{\frac{1}{2m}} + K_2 \varepsilon^{\sigma} \quad ,$$

taking into consideration that $|w_\varepsilon|$ is bounded and choosing K_1 first, then, provided that K_2, K_4 are sufficiently large, it follows that $\tilde{z}_\varepsilon$ majorizes the function z_ε.

From

$$\tilde{z}_\varepsilon \pm z_\varepsilon \ge 0 \quad \text{in } 0 < \rho < \rho_0, \qquad \tilde{z}_\varepsilon \pm z_\varepsilon\big|_\Gamma = 0$$

we obtain

$$\frac{\partial}{\partial\nu}\left(\tilde{z}_\varepsilon \pm z_\varepsilon\right) \leq 0$$

and, consequently,

$$\left|\frac{\partial z_\varepsilon}{\partial\nu}\right| \leq \left|\frac{\partial \tilde{z}_\varepsilon}{\partial\nu}\right| .$$

By means of $|\varepsilon^{\frac{1}{2m}} w_\varepsilon'| \leq K_m'$ this yields

$$\left|\frac{\partial z_\varepsilon}{\partial\nu}\right| \leq K_1 K_m' + K_2(\varepsilon) K_5 \varepsilon^{-\frac{1}{2m}} = \varepsilon^{-\frac{1}{2m}}\left((K_1 K_m' + K_4 K_5)\varepsilon^{\frac{1}{2m}} + K_2 K_5 \varepsilon^{\sigma}\right),$$

and finally

$$\lim_{\varepsilon\to 0} \varepsilon^{\frac{1}{l+1}}\left|\frac{\partial z_\varepsilon}{\partial\nu}\right| = 0 .$$

In the case of $l = 1$ the procedure is analogous, we have to choose only

$$K_2(\varepsilon) = K_4 \varepsilon^{\frac{1}{2}} |\ln \varepsilon| + K_2 \varepsilon^{\sigma}$$

and obtain

$$\left|\frac{\partial z_\varepsilon}{\partial\nu}\right| \leq K_1 K_2' + K_2(\varepsilon) K_5 \varepsilon^{-\frac{1}{2}} = \varepsilon^{-\frac{1}{2}}\left(K_1 K_2' \varepsilon^{\frac{1}{2}} + K_4 K_5 \varepsilon^{\frac{1}{2}} |\ln \varepsilon| + K_2 K_5 \varepsilon^{\sigma}\right).$$

If l is an even number with $l = 2m$, we set

$$\tilde{z}_\varepsilon = -K_1 \varepsilon^{\frac{1}{2m+1}} v_\varepsilon(\rho) + K_1 \varepsilon^{\frac{1}{2m+1}} v_\varepsilon\big|_{\rho=0} \tilde{v} + K_2(\varepsilon)(1-\tilde{v})$$

since $v_\varepsilon(\rho)$ does not vanish for $\rho = 0$.
From this it follows that

$$L_\varepsilon \tilde{z}_\varepsilon = K_1 \varepsilon^{\frac{2m}{2m+1}} - \left(-K_1 \varepsilon^{\frac{1}{2m+1}} v_\varepsilon\big|_{\rho=0} + K_2(\varepsilon)\right) L_\varepsilon \tilde{v}$$

and

$$\tilde{z}_\varepsilon\big|_{\rho=\rho_0} = -K_1 \varepsilon^{\frac{1}{2m+1}} v_\varepsilon\big|_{\rho=\rho_0} + K_2(\varepsilon) - \left(-K_1 \varepsilon^{\frac{1}{2m+1}} v_\varepsilon\big|_{\rho=0} + K_2(\varepsilon)\right) \tilde{v}\big|_{\rho=\rho_0} .$$

Because of the uniform boundedness of $v_\varepsilon(\rho)$ we obtain, as above,

$$\left|\frac{\partial z_\varepsilon}{\partial\nu}\right| \leq K_1 K_m' + K_1 K_5 + K_2(\varepsilon) K_5 \varepsilon^{-\frac{1}{2m+1}}$$

consequently

$$\lim_{\varepsilon\to 0} \varepsilon^{\frac{1}{l+1}} \left|\frac{\partial z_\varepsilon}{\partial \nu}\right| = 0 \quad . \qquad \blacksquare$$

Let n = 2, Ω the unit circle and

$$L_o \equiv (1-r^2)^l \left(x_1 \frac{\partial}{\partial x_1} + x_2 \frac{\partial}{\partial x_2}\right)$$

as well as $L_1 = -\Delta$.
Then the assumptions (H 6), (H 7), (H 8) are fulfilled with

$$P = \frac{(1-r^2)^{l+1}}{l+1}$$

and

$$B_o^* = -(1-\rho)(2-\rho)^l \ .$$

Since B_o^* does not depend on φ for $\rho = 0$, Theorem 4.17 yields

$$\lim_{\varepsilon\to 0} u_\varepsilon(0) = \frac{1}{2\pi}\int_0^{2\pi} g(\varphi)\, d\varphi \quad .$$

Now, instead of (H 8), we assume

(H 8)* There exists exactly one φ_κ such that $B_\varkappa(0,\varphi) \equiv 0$ and $B_k(0,\varphi) \neq 0$, $k \neq \varkappa$.
Furthermore, let

$$B_o(\rho,\varphi) = \rho\, B_o^*(\rho,\varphi) \qquad \text{with } B_o^* < 0 \quad \text{for } 0 < \rho < \rho_o \ .$$

Since for the corresponding boundary layer problem according to A 3, there exist solutions of classical boundary layer type we can transform Theorem 4.16 of this section to the case that (H 6) and (H 8)* are fulfilled.
Let n = 2, Ω be the unit circle and

$$L_o \equiv \left[x_1(1-r^2) - x_2\right] \frac{\partial}{\partial x_1} + \left[x_2(1-r^2) + x_1\right]\frac{\partial}{\partial x_2} \ .$$

Then (H 6) and (H 8)* are fulfilled.
It is not clear how to calculate $\lim_{\varepsilon\to 0} u_\varepsilon(0)$ in this case, because for the given example there does not exist a potential.

4.5.1.3. Closed characteristics in the neighbourhood of the boundary

Let n = 2 and let the behaviour of the characteristics in the neighbourhood of the boundary be characterised by (compare with (H 8), (H 8)*)

(H 9) $\quad B_o(\rho,\varphi) = \rho^l B_o^*(\rho,\varphi) \qquad$ with $B_o^* \neq 0$ and $l \geq 2$

and

$$B_1(\rho, \varphi)\Big|_{\rho=0} \neq 0 \quad .$$

A typical example in the case of the unit circle is the following

$$L_0 \equiv [x_1(1-r^2)^l - x_2]\frac{\partial}{\partial x_1} + [x_2(1-r^2)^l + x_1]\frac{\partial}{\partial x_2} \,, \qquad l \geq 2 \,.$$

For

$$L_0 \equiv - x_2 \frac{\partial}{\partial x_1} + x_1 \frac{\partial}{\partial x_2} \quad ,$$

that means actually $l = \infty$, we have closed characteristics. Since it is possible to investigate the case $l \geq 2$ analogously (for $l < \infty$ $\partial\Omega$ is then an instable limit cycle), we assume only $l \geq 2$ some in opposite to the headline.

In the neighbourhood of the boundary according to (H 9), we obtain the boundary layer problem

$$-A_{2,0}(0,\varphi)\frac{\partial^2 v}{\partial \zeta^2} + B_1(0,\varphi)\frac{\partial v}{\partial \varphi} = 0$$

$$v\Big|_{\zeta=0} = \tilde{g}(\varphi) \,.$$

Setting

$$\gamma(\varphi) = \frac{B_1(0,\varphi)}{A_{2,0}(0,\varphi)}$$

v_0 is of boundary layer type if and only if

$$\frac{1}{p}\int_0^{2\pi} \gamma(\varphi)\,\tilde{g}(\varphi)\,d\varphi = 0 \qquad with \qquad p = \int_0^{2\pi} \gamma(\varphi)\,d\varphi \,.$$

Since a constant u_0 is global approximation, we set $\tilde{g} = g - u_0$ and obtain

$$u_0 = \frac{1}{p}\int_0^{2\pi} \gamma(\varphi)\,g(\varphi)\,d\varphi \,.$$

$u_0 + \psi v_0$ satisfies

$$|L_\varepsilon(u_\varepsilon - (u_0 + \psi v_0))| \leq K\varepsilon^{1/2}$$

$$u_\varepsilon - (u_0 + \psi v_0)\Big|_\Gamma = 0 \,.$$

According to the construction we can even sharpen the first statement to

$$|L_\varepsilon(u_\varepsilon - (u_0 + \psi v_0))| \leq K\varepsilon^{\frac{1}{2}} \exp\left(-\frac{e\,\rho}{\sqrt{\varepsilon}}\right)\psi$$

(e is an arbitrary positive constant).

Let ψ have the properties $\psi \geq 0$, $\psi \equiv 1$ in $[0, \frac{\varrho_0}{3}]$, $\psi' \geq 0$, $\psi \equiv 0$ for $\varrho \geq \varrho_0$.

For

$$S_\varepsilon(x) = K_1\sqrt{\varepsilon}\left[K_2 - \exp\left(-\frac{e\varrho}{\sqrt{\varepsilon}}\right)\psi\right]$$

it follows that

$$L_\varepsilon S_\varepsilon \equiv 0 \qquad for \quad \varrho \geq \varrho_0$$

and otherwise

$$L_\varepsilon S_\varepsilon \geq K_1\sqrt{\varepsilon}\, e^{-\frac{e\varrho}{\sqrt{\varepsilon}}}\left[\psi(A_{2,0}e^2 + B_1\sqrt{\varepsilon}) + B_2\sqrt{\varepsilon}\,\psi' + \varepsilon B_3\psi''\right].$$

S_ε is barrier function for a corresponding choice of e, $K_{1,2}$, consequently, it holds

<u>Theorem 4.18.</u> Under the assumption (H 9) it holds that

$$\| u_\varepsilon - (u_0 + \psi v_0)\| \leq K\varepsilon^{1/2} \quad ,$$

in particular,

$$\lim_{\varepsilon\to 0} u_\varepsilon(x) = \frac{1}{P}\int_0^{2\pi} \gamma(\varphi)\, g(\varphi)\, d\varphi \qquad for \quad x \in \Omega \quad \blacksquare$$

<u>4.5.1.4. An absolute stable singular point with $\Gamma = \Gamma_-$</u>

We consider the model problem

$$L_\varepsilon u_\varepsilon \equiv \varepsilon L_1 u_\varepsilon - \left(x_1\frac{\partial u_\varepsilon}{\partial x_1} + x_2\frac{\partial u_\varepsilon}{\partial x_2}\right) + c\,u_\varepsilon = f \quad in\ \Omega = \{x \mid |x| < 1\}$$

$$u_\varepsilon|_\Gamma = g$$

where c denotes a constant.

Since it holds that $b \cdot \nu < 0$ on the boundary, it is the question whether u_ε converges uniformly to the solution of the global problem. It holds $\hat{\tau}_\varepsilon = 2$, but it is not to expect that, for $c > -2$, u_ε converges uniformly to u_0, since the inverse stability of L_ε(and L_0) can be guaranteed (in L_∞) only for $c > 0$.

Actually, it holds

$$\|u_\varepsilon - u_0\| \leq K \begin{cases} \varepsilon & c > 2 \\ \varepsilon^{\frac{c}{2}} & 0 < c < 2,\ c \neq 1 \\ \varepsilon^{\frac{c}{2}-\alpha} & c = 1,2, \quad \alpha > 0, arbitrary\ . \end{cases}$$

Out of a neighbourhood of (0,0) it actually holds $\lim_{\varepsilon\to 0} u_\varepsilon = u_o$ for $c > -2$, namely in $x^2 + y^2 \geqq 12\varepsilon$.
For a uniform approximation in the case $c = 0$ in $\overline{\Omega}$ we need a local approximation in the neighbourhood of (0,0).
Let $c = 0$ in the above example and

$$u = -\sum_{1\leq\beta_1+\beta_2\leq 2} \frac{f_{\beta_1\beta_2} x_1^{\beta_1} x_2^{\beta_2}}{\beta_1+\beta_2} - \int_0^1 f^*(x_1\tau, x_2\tau)\tau^{-1}\,d\tau$$

where f^* denotes the remainder of the corresponding Taylors'series of f at (0,0).
Then it holds

$$L_o u = f - f_{oo} .$$

Let $\quad \varrho = \frac{(x_1^2+x_2^2)^{1/2}}{\varepsilon^{1/2}} = \frac{r}{\varepsilon^{1/2}}$, let $v(\varrho)$ satisfy

$$v''(\varrho) + (\varrho + \tfrac{1}{\varrho})v'(\varrho) = f_{oo}$$

$$v(\varepsilon^{-\frac{1}{2}}) = 0$$

and let $w_\varepsilon(r,\varphi)$ be the solution of

$$\varepsilon\,\Delta w_\varepsilon + r\frac{\partial w_\varepsilon}{\partial r} = 0$$
$$w_\varepsilon\big|_{r=1} = g(\varphi) - u\big|_{r=1} .$$

Then it is possible to express w_ε by means of confluent hypergeometric functions in the form

$$w_\varepsilon = \sum_{\mu=-\infty}^{\infty} d_\mu f_\mu r^{|\mu|} e^{i\mu\varphi}\, {}_1F_1(\tfrac{1}{2}|\mu|, |\mu|+1, -\tfrac{r^2}{2\varepsilon}) \quad .$$

It holds that

$$|L_\varepsilon(u_\varepsilon - (u+v+w_\varepsilon))| \leq K\begin{cases} \varepsilon^{1/2} & \text{for } r \leq \varepsilon^{1/2} \\ \frac{\varepsilon}{r} & \text{for } r \geq \varepsilon^{1/2} \end{cases}$$

and the estimate

$$\| u_\varepsilon - (u+v+w_\varepsilon)\| \leq K\begin{cases} \varepsilon^{1/2} & \text{for } r \leq \varepsilon^{1/2} \\ \frac{\varepsilon}{r} & \text{for } r \geq \varepsilon^{1/2} \end{cases}$$

has been shown by DE GROEN (1974/75) by means of the barrier function

$$S_\varepsilon = \varepsilon \frac{{}_1F_1(\frac{1}{2},1,-\frac{r^2}{2\varepsilon})}{{}_1F_1(\frac{1}{2},1,-\frac{1}{2\varepsilon})} .$$

<u>4.5.2. Singular lines</u>

We consider elliptic boundary value problems of the form

$$L_\varepsilon u_\varepsilon \equiv -\varepsilon L_1 u_\varepsilon + L_0 u_\varepsilon = 0 \qquad \text{in } \Omega \tag{4.20}$$
$$u_\varepsilon|_\Gamma = g$$

with

$$L_1 \equiv \sum a_{ij} \frac{\partial^2}{\partial x_i \partial x_j} + \sum a_i \frac{\partial}{\partial x_i} + a_0 , \qquad a_0 \le 0$$

$$L_0 \equiv b_1 \frac{\partial}{\partial x_1} + b_2 \frac{\partial}{\partial x_2} .$$

Let Ω be a two-dimensional domain discribed by

$$\Omega = \{ (x_1,x_2) \mid -1 < x_1 < 1 , -f_1(x_1) < x_2 < f_2(x_1) \}$$

or

$$\Omega = \{ (x_1,x_2) \mid -1 < x_1 < 1 , \ 0 < x_2 < f_2(x_1) \}$$

by means of regular functions $f_{1,2}(x_1)$.
Then Ω is an admissible domain in the sense of Chapter 2.
We assume

(H 10) Let all points at $x_2 = 0$ and only these be stationary points of $b(x)$. Then $x_2 = 0$ is called a <u>singular line</u>.

We study the special cases in which the characteristics are parallel or perpendicular to the singular line.

Fig. 4.2

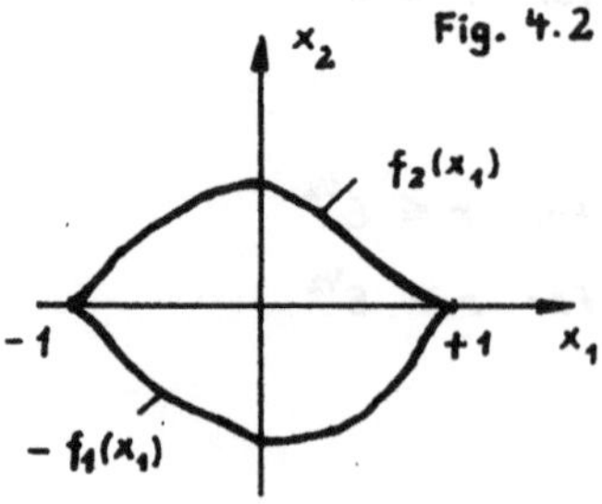

Fig. 4.3

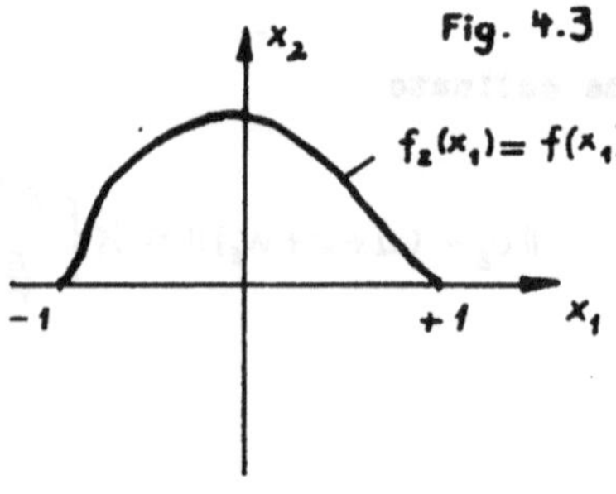

4.5.2.1. A singular line parallel to the characteristics

We consider

$$L_\varepsilon u_\varepsilon \equiv -\varepsilon L_1 u_\varepsilon + x_2^l \frac{\partial u_\varepsilon}{\partial x_1} = 0 \quad in\ \Omega$$

$$u_\varepsilon\big|_\Gamma = g$$

with

$$\Omega = \{ (x_1, x_2) \mid \ -1 < x_1 < 1\ , \quad 0 < x_2 < f(x_1) \}$$

and assume that the lines x_2 = const. intersect $\partial\Omega$ at exactly two points (except $x_2 \leq 0$ and $x_2 \geq \sup f(x_1)$). Without restriction of generality let $0 = f(\pm 1) < f(x_1) < f(0)$ for $0 < |x_1| < 1$ such that

$$\Gamma_- = \{ (x_1, x_2) \in \Gamma \mid \ -1 \leq x_1 < 0\ , \quad x_2 = f(x_1) \}$$
$$\Gamma_+ = \{ (x_1, x_2) \in \Gamma \mid \ 0 < x_1 \leq 1\ , \quad x_2 = f(x_1) \}\ .$$
(cf. Fig. 4.3)

Let u be the solution of $\frac{\partial u}{\partial x_1} = 0$, $u\big|_{\Gamma_-} = g$. u is regular in Ω_γ^o with $\Omega_\gamma = \Omega \cap \{ x \mid \text{dist}\ (x, \bar{x}) > \gamma \}$ where $\bar{x} = (0, f(0))$.
In the neighbourhood of the singular line $x_2 = 0$ let

$$\zeta = \frac{x_2}{\varepsilon^{\frac{1}{l+2}}}$$

and let v satisfy

$$\frac{\partial^2 v}{\partial \zeta^2} - \zeta^l \frac{\partial v}{\partial x_1} = 0$$
$$v\big|_{x_2=0} = (g-u)\big|_{x_2=0}\ .$$

For this boundary layer problem there exists a solution of classical boundary layer type (cf. A 4.1), because of the singularity of

$$\frac{\partial^2 v}{\partial x_1 \partial y}\ ,\ \frac{\partial^2 v}{\partial x_1^2} \quad in\ (-1,0)$$

a regularization is applied.

<u>Theorem 4.19.</u> Let $\Omega_\gamma = \Omega \cap \{ x \mid \text{dist}\ (x, \Gamma_+) > \gamma \}$. Then it holds

$$\| u_\varepsilon - (u+v) \| \leq K\, \varepsilon^{\frac{1}{2(2+l)}} \qquad in\ \Omega_\gamma\ . \quad \blacksquare$$

Proof. In $\Omega_{\frac{\gamma}{2}}$, for corresponding regularisation, it holds that

$$|L_\varepsilon(u_\varepsilon-(u+v))|\leq K\varepsilon^{1-\frac{3}{2(l+2)}}$$

$$|u_\varepsilon-(u+v)|\Big|_{x_2=0}\leq K\varepsilon^{\frac{1}{2(l+2)}}$$

$$u_\varepsilon-(u+v)\Big|_{\Gamma}=0$$

$$|u_\varepsilon-(u+v)|\leq K$$

according to the construction. Now we apply Theorem 2.10, case (vi). Suppose that $u_\varepsilon-(u+v)$ is bounded by $K\varepsilon^{\frac{1}{2(l+2)}}$ on the whole boundary, according to Theorem 2.10 we obtain

$$|u_\varepsilon-(u+v)|\leq K\varepsilon^{-\varkappa}\,\varepsilon^{1-\frac{3}{2(l+2)}}+K\varepsilon^{\frac{1}{2(l+2)}}$$

with $\varkappa=\frac{1}{2+l}$ and this would be the assertion.
Now let $\tilde{V}_\varepsilon$ be the barrier function corresponding to Theorem 2.10 and

$$S=\tilde{V}_\varepsilon+K_1\psi(x_1,x_2)$$

with $\psi(x_1,x_2)=0$ in Ω_γ, $\psi\equiv 1$ for $\operatorname{dist}(x,\Gamma_+)<\frac{\gamma}{2}$, otherwise let ψ be regular. For a corresponding choice of K_1 then $u_\varepsilon-(u+v)$ is majorized by S in $\Omega_{\frac{\gamma}{2}}$. The restriction to Ω_γ yields the assertion■

Remark. Of course, it is possible to construct a local corrector in the neighbourhood of Γ_+, whereas the procedure in the neighbourhood of $(0,f(0))$ is analogous to that in 4.3. The parabolic boundary layer coincides with the ordinary boundary layer. The treatment of such problems can be found in IL'IN/LELIKOVA (1975)■

The case $\Omega=\{(x_1,x_2)\,|\,-1<x_1<1,\ -f_1(x_1)<x_2<f_2(x_1)$ is not interesting, since nothing happens along the singular line.

4.5.2.2. A singular line perpendicular to the characteristics

We consider

$$L_\varepsilon u_\varepsilon\equiv-\varepsilon L_1u_\varepsilon\pm x_2^l\frac{\partial u_\varepsilon}{\partial x_2}=0\qquad\text{in }\Omega$$

$$u_\varepsilon\Big|_\Gamma=g$$

with

$$\Omega=\{(x_1,x_2)\,|\,-1<x_1<1,\ -f_1(x_1)<x_2<f_2(x_1)\}\quad.\text{(cf. Fig. 4.2)}$$

Let $f_{1,2}$ be sufficiently regular and $f_1(1) = f_2(1) = f_1(-1) = f_2(-1) = 0$. Furthermore, let $f_{1,2}$ vanish at $x = \pm 1$ of the order $m^{\pm}_{1,2}$. Finally, let

$$g\Big|_{x_2=f_2(x_1)} = g_2(x_1)\,, \qquad g\Big|_{x_2=-f_1(x_1)} = g_1(x_1)$$

and l be even.

Then, without restriction of generality we can assume that $L_0 = -x_2^{2k}\frac{\partial}{\partial x_2}$ and it holds

$$\Gamma_- = \{(x_1,x_2)\,|\ -1 < x_1 < 1,\ x_2 = f_2(x_1)\},\quad \Gamma_+ = \{(x_1,x_2)\,|\ -1 < x_1 < 1,\ x_2 = -f_1(x_1)\}\,.$$

Consequently, the solution of the global problem becomes $g_2(x_1)$. In the neighbourhood of Γ_+ let $\zeta = \dfrac{x_2+f_1(x_1)}{\varepsilon}$ and

$$L_\varepsilon = \varepsilon^{-1}\Big[-a(x_1)\frac{\partial^2}{\partial\zeta^2} - f_1^l(x_1)\frac{\partial}{\partial\zeta}\Big] + L_\varepsilon^* \qquad .$$

If $v = v_0 + \varepsilon v_1$ is a local corrector with $|L_\varepsilon v| \le K\varepsilon$, such that the boundary conditions on Γ_+ are satisfied by $v + g_2$, then we obtain

$$v_0 = [\,g_1(x_1) - g_2(x_1)]\exp\Big(-\frac{f_1^l}{a}\,\frac{f_1(x_1)+x_2}{\varepsilon}\Big)\,.$$

On Γ_- v_0 satisfies

$$v_0\Big|_{\Gamma_-} = [g_1(x_1) - g_2(x_1)]\exp\Big(-\frac{f_1^l}{a}\,\frac{f_1+f_2}{\varepsilon}\Big)$$

and is exponentially small in the domain $|x_1| \le x_0 < 1$.

In the neighbourhood of, for instance, $x_1 = 1$ $v_0\big|_{\Gamma_-}$ behaves as

$$\mu\exp\Big(-\frac{\mu^{m_1^+ l + \min(m_1^+, m_2^+)}}{\varepsilon}\Big)\quad .$$

Since for arbitrary positive r, it holds

$$\mu\exp\Big(-\frac{\mu^r}{\varepsilon}\Big) \le K\varepsilon^{\frac{1}{r}} \qquad ,$$

which implies

$$\Big|\,v_0\big|_{\Gamma_-}\Big| \le K\varepsilon^{\frac{1}{s_1}} \qquad ,$$

where $s_1 = \max(\min(m_1^+, m_2^+) + 2k\, m_1^+,\ \min(m_1^-, m_2^-) + 2k\, m_1^-)$.

Summing up we obtain

$$|L_\varepsilon(u_\varepsilon - (g_2 + v))| \le K\varepsilon$$

$$|u_\varepsilon - (g_2+v)|_\Gamma \le K\varepsilon^{\frac{1}{s_1}} \quad .$$

The barrier function

$$K_1 \varepsilon^{\frac{1}{s_1}} - K_2 \varepsilon^{\frac{2}{2k+1}} V_\varepsilon(x_2)$$

(cf. Theorem 2.10, (iv)) yields the estimate $|u_\varepsilon-(g_2+v)| \le K\varepsilon^{\frac{2}{2k+1}} + K\varepsilon^{\frac{1}{s_1}}$.
Because of $s_1 \ge 2k+1$ we obtain

<u>Theorem 4.20c.</u> Let $L_0 = -x_2^{2k}\frac{\partial}{\partial x_2}$ and let the singular line lie in the interior of Ω.
Then it holds that

$$\| u_\varepsilon - (g(x_1) + v_0) \| \le K\varepsilon^{\frac{1}{s_1}}$$

with $\quad s_1 = max[\, min(m_1^+, m_2^+) + 2k\, m_1^+ ,\ min(m_1^-, m_2^-) + 2k\, m_1^-]$. ■

Considering the subdomains $x_2 > 0$ and $x_2 < 0$, respectively, we obtain analogous problems.
First of all let

$$\Omega = \{ (x_1, x_2) \mid -1 < x_1 < 1,\ 0 < x_2 < f_2(x_1) \} .$$

Again $g_2(x_1)$ is the solution of the global problem. Setting

$$\zeta = \frac{x_2}{\varepsilon^{\frac{1}{2k+1}}}$$

it holds

$$L_\varepsilon = \varepsilon^{\frac{2k-1}{2k+1}} \left[-a(x_1)\frac{\partial^2}{\partial\zeta^2} - \zeta^{2k}\frac{\partial}{\partial\zeta} \right] + \varepsilon^{\frac{2k}{2k+1}} L_\varepsilon^* .$$

Requiring

$$a\frac{\partial^2 v}{\partial\zeta^2} + \zeta^{2k}\frac{\partial v}{\partial\zeta} = 0, \qquad v\Big|_{\zeta=0} = g_1(x_1) - g_2(x_1) ,$$

we obtain

$$v = [\, g_1(x_1) - g_2(x_1)]\,\frac{1}{A}\int_{\infty}^{\frac{x_2}{\varepsilon^{\frac{1}{2k+1}}}} exp\left(-\frac{1}{a(2k+1)}t^{2k+1}\right)dt .$$

Now $u_\varepsilon - (g_2(x_1) + v)$ satisfies

$$|L_\varepsilon(u_\varepsilon - (g_2(x_1)+v))| \le K\varepsilon^{\frac{2k}{2k+1}}$$

$$u_\varepsilon - (g_2(x_1)+v)\Big|_{x_2=0} = 0$$

$$|u_\varepsilon - (g_2(x_1)+v)\Big|_{x_2=f_2(x_1)}| \le K\varepsilon^{\frac{1}{s_2}}, \qquad s_2 = (2k+1)\,max(m_2^+, m_2^-) ,$$

where the last follows from the estimate

$$\mu \left| \int_{\infty}^{\frac{\mu^m}{\varepsilon^{\frac{1}{2k+1}}}} \exp\left(-\frac{t^{2k+1}}{2k+1}\right) dt \right| \leq K^* \varepsilon^{\frac{1}{m(2k+1)}} .$$

Then, because of $s_2 \geq 2k+1$, the barrier function (cf. Chapter 2)

$$-K_1 \varepsilon^{\frac{1}{2k+1}} V_\varepsilon(x_2) + K_2 \varepsilon^{\frac{1}{s_2}}$$

yields

<u>Theorem 4.20b.</u> Let $L_0 = -x_2^{2k} \frac{\partial}{\partial x_2}$ and let the singular line belong to the boundary of Ω. Then it holds

$$\| u_\varepsilon - (g_2(x_1) + v) \| \leq K \varepsilon^{\frac{1}{s_2}}$$

with

$$s_2 = (2k+1) \max(m_2^+, m_2^-) . \qquad \blacksquare$$

The case

$$\Omega = \{ (x_1, x_2) \mid -1 < x_1 < 1 , 0 > x_2 > -f_1(x_1) \}$$

is analogous to that which has been considered first.

Now let l be odd, $l = 2k-1$.
In this case the sign of $\pm x_2^{2k-1} \frac{\partial}{\partial x_2}$ is of decisive importance. First we consider

$$L_\varepsilon u_\varepsilon \equiv -\varepsilon L_1 u_\varepsilon - x_2^{2k-1} \frac{\partial u_\varepsilon}{\partial x_2} = 0 \qquad \text{in } \Omega$$
$$u_\varepsilon \big|_\Gamma = g$$

in

$$\Omega = \{ (x_1, x_2) \mid -1 < x_1 < 1 , -f_1(x_1) < x_2 < f_2(x_1) \} .$$

The solution of the global problem becomes

$$u = \begin{cases} g_2(x_1) , & x_2 > 0 \\ g_1(x_1) , & x_2 < 0 \end{cases}$$

and at $x_2 = 0$ a free boundary layer originates. By setting

$$\zeta = \frac{x_2}{\varepsilon^{\frac{1}{2k}}}$$

in the neighbourhood of $x_2 = 0$ we obtain

$$L_\varepsilon = \varepsilon^{\frac{2k-2}{2k}}\left[-a(x_1)\frac{\partial^2}{\partial\zeta^2} - \zeta^{2k-1}\frac{\partial}{\partial\zeta}\right] + \varepsilon^{\frac{2k-1}{2k}} L_\varepsilon^* .$$

The requirement

$$a\frac{\partial^2 v}{\partial\zeta^2} + \zeta^{2k-1}\frac{\partial v}{\partial\zeta} = 0, \quad v\Big|_{\zeta=-\infty} = \frac{g_1-g_2}{2}, \quad v\Big|_{\zeta=+\infty} = \frac{g_2-g_1}{2}$$

yields

$$v = \frac{g_2-g_1}{2A^*}\int_0^\zeta \exp\left(-\frac{1}{a2k}t^{2k}\right)dt \qquad \text{with } A^* = \int_0^\infty \exp\left(-\frac{1}{a2k}t^{2k}\right)dt .$$

__Theorem 4.20c.__ Let $L_0 \equiv -x_2^{2k-1}\frac{\partial}{\partial x_2}$ and let the singular line lie in the interior of Ω. Then it holds that

$$\|u_\varepsilon - \left(\frac{g_2+g_1}{2} + v\right)\| \leq K\begin{cases} \varepsilon^{1/2}\ln\varepsilon & \text{if } k=1 \text{ and } s_3 = 1 \\ \varepsilon^{\frac{1}{s_3}} & \text{otherwise} \end{cases}$$

with

$$s_3 = 2k\max(m_2^+, m_2^-, m_1^+, m_1^-) . \qquad \blacksquare$$

__Proof.__ According to the construction it holds

$$\left|L_\varepsilon\left(u_\varepsilon - \left(\frac{g_2+g_1}{2} + v\right)\right)\right| \leq K\varepsilon^{\frac{2k-1}{2k}} .$$

Since v is a classical boundary layer function, we obtain

$$\frac{g_2+g_1}{2} + v\Big|_\Gamma = g + r \quad \text{with } |r| \leq K\varepsilon^N \text{ and arbitrary } N$$

in the domain $|x| < x_0 < 1$. Analogous to the above we obtain, however, only

$$\left|u_\varepsilon - \left(\frac{g_2+g_1}{2} + v\right)\right|_\Gamma \leq K\varepsilon^{\frac{1}{s_3}} \qquad \text{with } s_3 = 2k\max(m_2^+, m_2^-, m_1^+, m_1^-) .$$

The a-priori estimate stated in Theorem 2.10 yields the assertion $\blacksquare$

In the case $\Omega = \{(x_1,x_2) \mid -1 < x_1 < 1,\ 0 < x_2 < f_2(x_1)$ we have a "usual" boundary layer of the above type and it does not change so much $\blacksquare$

Now we consider

$$L_\varepsilon u_\varepsilon \equiv -\varepsilon L_1 u_\varepsilon + x_2^{2k-1}\frac{\partial u_\varepsilon}{\partial x_2} = 0 \qquad \text{in } \Omega$$
$$u_\varepsilon\big|_\Gamma = g$$

in $\Omega = \{(x_1,x_2) \mid -1 < x_1 < 1,\ -f_1(x_1) < x_2 < f_2(x_1)\}$.

In this case we have

$$\Gamma_+ = \Gamma \setminus \{ (1,0) \cup (-1,0) \} .$$

Therefore, it is not possible to impose conditions on the solution of the global problem $U(x_1)$. In difference to the cases which have been considered till now the domain of influence of Γ does not correspond to Ω, here it is even empty. Let $v^{1,2}$ be the local approximations in the neighbourhood of the boundary with $|L_\varepsilon v^{1,2}| \leq K\varepsilon^2$, such that $v^{1,2} + U$ satisfies the boundary conditions for $x_2 = f_2(x_1)$ and $x_2 = -f_1(x_1)$, respectively. Then it holds $v^{1,2} = v_0^{1,2} + \varepsilon v_1^{1,2} + \varepsilon^2 v_2^{1,2}$ with

$$v_0^1 = (g_1 - U) \exp\left(- \frac{f_1^{2K-1}}{a} \frac{f_1(x_1)+x_2}{\varepsilon}\right)$$

$$v_0^2 = (g_2 - U) \exp\left(- \frac{f_2^{2K-1}}{a} \frac{f_2(x_1)-x_2}{\varepsilon}\right) .$$

Let

$$L_1 \equiv a_{11} \frac{\partial^2}{\partial x_1^2} + a_1 \frac{\partial}{\partial x_1} + a_0 + L_1^* .$$

Obviously, it holds

$$L_\varepsilon U = \varepsilon \left[a_{11} \frac{\partial^2 U}{\partial x_1^2} + a_1 \frac{\partial U}{\partial x_1} + a_0 U\right] ,$$

that means, if a_{11}, a_1 and a_0 do not depend on x_2, then we can improve the accuracy of the approximation by requiring

$$a_{11} \frac{\partial^2 U}{\partial x_1^2} + a_1 \frac{\partial U}{\partial x_1} + a_0 U = 0 .$$

Setting

$$\phi_\varepsilon = U(x) + v^1 + v^2$$

it follows

$$|L_\varepsilon \phi_\varepsilon| \leq K\varepsilon^2 .$$

In the domain $|x_1| \leq x_0 < 1$ v_0^1 and v_0^2 are arbitrary small for $x_2 = f_2(x_1)$ and $x_2 = -f_1(x_1)$, respectively. In the neighbourhood of $|x_1| = 1$ we obtain the conditions

$$U(1) = g(1) \quad \text{and} \quad U(-1) = g(-1)$$

from the requirements

$$|v_0^1|_{x_2=f_2(x_1)}| \leq K\varepsilon^{\frac{1}{s}} , \qquad |v_0^2|_{x_2=-f_1(x_1)}| \leq K\varepsilon^{\frac{1}{s}} , \qquad s>0 .$$

Since $a_0 \leq 0$, the boundary value problem

$$a_{11}\frac{\partial^2 U}{\partial x_1^2} + a_1 \frac{\partial U}{\partial x_1} + a_0 U = 0$$
$$U(1) = g(1)\,, U(-1) = g(-1)$$

has a uniquely determined solution.
For the s we obtain, as above, that

$$s = s_4 = max\left[min(m_1^+, m_2^+) + l m_1^+, min(m_1^+, m_2^+) + l m_2^+, min(m_1^-, m_2^-) + l m_1^-, min(m_1^-, m_2^-) + l m_2^-\right].$$

The barrier function is based on case (v) in Theorem 2.10, because $b_1 \equiv 0$. Setting

$$S^* = K_1 \varepsilon^{\frac{1}{s_4}} + K_2 S_\varepsilon(x)$$

we immediately obtain

<u>Theorem 4.21.</u> Let $L_0 \equiv x_2^{2k-1}\frac{\partial}{\partial x_2}$ and let the singular line lie in the interior of the domain Ω. Furthermore, let a_{11}, a_1, a_0 be independent of x_2. Then it holds

$$\| u_\varepsilon - \left(U(x_1) + v_0^1 + v_0^2 \right) \| \leq K \varepsilon^{\frac{1}{s_4}}$$

with the above defined s_4 where $U(x_1)$ denotes the solution of the boundary value problem

$$a_{11} U'' + a_1 U' + a_0 U = 0$$
$$U(1) = g(1),\ U(-1) = g(-1) \quad . \quad \blacksquare$$

a_{11}, a_1 and a_0 can depend on x_2 in the case $k = 1$. Then, we require

$$a_{11}(x_1, 0)\frac{\partial^2 U}{\partial x_1^2} + a_1(x_1, 0)\frac{\partial U}{\partial x_1} + a_0(x_1, 0) U = 0 \quad ,$$

set

$$\phi_\varepsilon = U(x_1) + \varepsilon U_1(x_1, x_2) + v^1 + v^2$$

and determine U_1 from

$$x_2 \frac{\partial U_1}{\partial x_2} = \left[a_{11}(x_1,x_2) - a_{11}(x_1,0)\right]\frac{\partial^2 U}{\partial x_1^2} + \left[a_1(x_1,x_2) - a_1(x_1,0)\right]\frac{\partial U}{\partial x_1} + \left[a_0(x_1,x_2) - a_0(x_1,0)\right] U \quad .$$

It always holds $s \geq 2$. $s = 2$ holds if and only if $m_1^+ = m_2^+ = m_1^- = m_2^- = 1$ and $k = 1$ (the assertion of GRASMAN (1977), that $s = 1$ for $k = 1$, is not valid).
The cases $\Omega = \{(x_1, x_2) \mid 0 < x_2 < f_2(x_1)\}$ and $\Omega = \{(x_1, x_2) \mid -f_1(x_1) < x_2 < 0\}$, respectively, are less interesting. There $g_1(x_1)$ and $g_2(x_1)$, respectively, is the solution of the reduced problem.

4.6. Nonlinear elliptic problems of second order

We consider quasilinear elliptic value problems of the form

$$A_\varepsilon u_\varepsilon \equiv \varepsilon L u_\varepsilon + \sum_{\mu=1}^{n} b_\mu(x,u_\varepsilon)\frac{\partial u_\varepsilon}{\partial x_\mu} + c(x,u_\varepsilon) = 0 \quad in\ \Omega \qquad u_\varepsilon = 0 \quad on\ \partial\Omega \tag{4.21}$$

in a bounded domain $\Omega \subset \mathbb{R}$ with regular boundary $\partial\Omega$. L denotes a linear, uniformly elliptic differential operator of order two. Let the functions b_μ, c and the coefficients of L be sufficiently regular.
In difference to the linear problems considered in 4.1 - 4.5 the uniqueness of a solution u_ε of (4.21) is not a-priori guaranteed. Applying Banach's fixed point theorem (see Chapter 2) we shall prove, that in a neighbourhood of a formal asymptotic solution ϕ_ε a uniquely determined solution u_ε of (4.21) does exist for which ϕ_ε is a uniform asymptotic solution.
As in the linear case regular solutions u_o of the reduced problem

$$\sum_{\mu=1}^{n} b_\mu(x,u)\frac{\partial u}{\partial x_\mu} + c(x,u) = 0 \qquad in\ \Omega \tag{4.22}$$

are of importance for the construction of formal asymptotic solutions. Now, the behaviour of the characteristics near the boundary additionally depends on the solution u_o. Analogous to the linear case we set $b(x,u) = (b_\mu(x,u))$, ν for the exterior unit normal and

$$\Gamma_\pm(u) = \{ x \in \partial\Omega \mid b(x,u(x))\cdot\nu(x) \gtrless 0 \} .$$

Our principal assumption in the following sections is the existence of a regular solution u_o of (4.22) for which it holds that $\partial\Omega = \Gamma_+(u_o)$ or $\partial\Omega = \Gamma_-(u_o)$ and $u_o|_{\partial\Omega} = 0$.

4.6.1. Problems without boundary layers

We assume

(H 11) There is a regular solution u_o of (4.22) with $u_o|_{\partial\Omega} = 0$ and $\partial\Omega = \Gamma_-(u_o)$.

Expanding the functions $b_\mu(x,.)$ and $c(x,.)$ at $u = u_o(x)$ in a Taylor's series for the global formal solution

$$u^N(x,\varepsilon) = \sum_{\mu=0}^{N} \varepsilon^\mu u_\mu(x)$$

we obtain problems of the form

$$\sum_{\mu=1}^{n} b_\mu(x,u_o(x))\frac{\partial u_\kappa}{\partial x_\mu} + c^*(x)u_\kappa = f_\kappa(x,u_o,\dots,u_{\kappa-1}) \qquad u_\kappa = 0 \quad on\ \partial\Omega \qquad \kappa = 1,\dots,N .$$

According to Chapter 3 under the condition

(H 12) $$c^*(x) \equiv \sum_{\mu=1}^{n} \frac{\partial b_\mu}{\partial u}(x,u_0(x)) \frac{\partial u_0}{\partial x_\mu} + \frac{\partial c}{\partial u}(x,u_0(x)) \geq c_0 > 0$$

c_0 sufficiently large

and (H 11) there exist uniquely determined regular solutions $u_1,\dots,u_N$. Setting $\phi_\varepsilon^N(x) = u^N(x,\varepsilon)$ we obtain

Theorem 4.22. Let (H 11) and (H 12) be fulfilled. Then there are constants K_1 and K_2 independent of ε such that there exists a solution u_ε of (4.21) uniquely determined in

$$B(\phi_\varepsilon^N, K_1\varepsilon) = \{ u_\varepsilon \in C_0^{2+\alpha} \mid |u_\varepsilon - \phi_\varepsilon^N|_{2+\alpha} \leq K_1\varepsilon \} .$$

ϕ_ε^N is a uniform asymptotic solution of (4.21) with

$$|u_\varepsilon - \phi_\varepsilon^N|_{2+\alpha} \leq K_2 \varepsilon^{N+1} . \qquad \blacksquare$$

Proof. For $z_\varepsilon = u_\varepsilon - \phi_\varepsilon^N$ we obtain the equivalent problem

$$A_\varepsilon(z_\varepsilon + \phi_\varepsilon^N) - A_\varepsilon \phi_\varepsilon^N = \tilde{L}_\varepsilon z_\varepsilon + R_\varepsilon z_\varepsilon = F_\varepsilon$$

with $[F_\varepsilon]_0 \leq K\varepsilon^{N+1}$, the linear operator

$$\tilde{L}_\varepsilon z \equiv \varepsilon L z + \sum_{\mu=1}^{n} b_\mu(x,\phi_\varepsilon^N) \frac{\partial z}{\partial x_\mu} + \Big[\sum_{\mu=1}^{n} \frac{\partial b_\mu}{\partial u}(x,\phi_\varepsilon^N) \frac{\partial \phi_\varepsilon^N}{\partial x_\mu} + \frac{\partial c}{\partial u}(x,\phi_\varepsilon^N) \Big] z$$

and the nonlinear operator

$$R_\varepsilon z \equiv \sum_{\mu=1}^{n-1} [b_\mu(x,\phi_\varepsilon^N + z) - b_\mu(x,\phi_\varepsilon^N)] \frac{\partial z}{\partial x_\mu} + \sum_{\mu=1}^{n} \Big[b_\mu(x,\phi_\varepsilon^N + z) - b_\mu(x,\phi_\varepsilon^N) - \frac{\partial b_\mu}{\partial u}(x,\phi_\varepsilon^N) z \Big] \frac{\partial \phi_\varepsilon^N}{\partial x_\mu} +$$
$$+ c(x,\phi_\varepsilon^N + z) - c(x,\phi_\varepsilon^N) - \frac{\partial c}{\partial u}(x,\phi_\varepsilon^N) z .$$

We now show that the assumptions of the Theorems 2.13 and 2.14 are fulfilled. The coefficients of $\tilde{L}_\varepsilon$ and their derivatives are uniformly bounded with respect to ε and from Theorem 2.14 we have

$$|z|_{2+\alpha} \leq K \left(\varepsilon^{1+\alpha} \|\tilde{L}_\varepsilon z\|_\alpha + [z]_0 \right) .$$

Because $\left| \sum_{\mu=1}^{n} \frac{\partial b_\mu}{\partial u}(x,\phi_\varepsilon^N) \frac{\partial \phi_\varepsilon^N}{\partial x_\mu} + \frac{\partial c}{\partial u}(x,\phi_\varepsilon^N) - c^*(x) \right| \leq K\varepsilon$ and

$c^*(x) \geq c_0 > 0$, already $\frac{2}{c_0}[\tilde{L}_\varepsilon z]_0$ majorizes z and from the maximum principle it follows

$$[z]_0 \leq \frac{2}{c_0}[\tilde{L}_\varepsilon z]_0 \qquad \text{for all } z \in C_0^{2+\alpha}(\bar{\Omega}) .$$

Choosing the ε-dependent norm

$$|u|_\alpha = [u]_0 + \varepsilon^{1+\alpha}[u]_\alpha$$

in the image space $C^\alpha(\bar{\Omega})$ we see that (H 10) from Chapter 2 is fulfilled by setting $X = C_0^{2+\alpha}(\bar{\Omega})$, $Y = C^\alpha(\bar{\Omega})$ with a ε-independent constant l.
We now investigate whether (H 11) from Chapter 2 is fulfilled. For this purpose we expand $b_\mu(x,.)$ and $c(x,.)$ at ϕ_ε in a Taylor series and obtain

$$|R_\varepsilon z_1 - R_\varepsilon z_2|_\alpha \leq \frac{K\sigma}{\varepsilon}|z_1 - z_2|_{2+\alpha}$$

for $0 \leq \sigma \leq \sigma^*$, $|z_1|_{2+\alpha} \leq \sigma$, $|z_2|_{2+\alpha} \leq \sigma$ such that (H 11) is fulfilled with $m(\sigma) = \frac{K\sigma}{\varepsilon}$. Further, it holds $|F_\varepsilon|_\alpha \leq K^*\varepsilon^{N+1}$, consequently for sufficiently small ε also $|F_\varepsilon|_\alpha \leq \frac{\sigma_0 l}{2} = \frac{\varepsilon l^2}{4K}$ ($N \geq 1$). Now the assertion of Theorem 4.22 follows from Theorem 2.13. ∎

<u>Remark.</u> The existence of a regular solution u_0 of (4.22) can be guaranteed according to ROMASKEVIČ (1981), and (H 11) as well as (H 12) are fulfilled if

$$\frac{\partial c}{\partial u}(x,u) \geq c_1 > 0 , \qquad c_1 \text{ sufficiently large}$$

on compact subsets M of $\bar{\Omega} \times R$ and $\partial\Omega = \Gamma_-(u)$ for all u with $(x,u(x)) \in M$. ∎

<u>4.6.2. Problems with ordinary boundary layers</u>

In difference to (H 11) we now assume

(H 13) There is a regular solution u_0 of (4.22) with $\partial\Omega = \Gamma_+(u_0)$.

For the global formal solution

$$u^N(x,\varepsilon) = \sum_{\mu=0}^{N} \varepsilon^\mu u_\mu(x)$$

we obtain problems of the form

$$\sum_{\mu=1}^{n} b_\mu(x,u_0(x)) \frac{\partial u_\kappa}{\partial x_\mu} + c^*(x)u_\kappa = f_\kappa(x,u_0,\ldots,u_{\kappa-1}), \qquad \kappa = 1,\ldots,N .$$

According to Chapter 3 (H 12) and (H 13) again guarantee uniquely determined regular solutions $u_1,\ldots,u_N$. The global formal solution does not satisfy the boundary condition in general, therefore local correctors in the neighbourhood of $\Gamma_+(u_0)$ are needed. The corrector $v(x,\varepsilon) = u_\varepsilon(x) - u^N(x,\varepsilon)$ satisfies the problem

$$\varepsilon L v + \sum_{\mu=1}^{n} B_\mu(x,v,\varepsilon)\frac{\partial v}{\partial x_\mu} + C(x,v,\varepsilon) = D(x,\varepsilon) \qquad (4.23)$$

$$v\big|_{\partial\Omega} = -u^N\big|_{\partial\Omega}$$

with the abbreviations

$$B_\mu(x,v,\varepsilon) = b_\mu(x,u^N(x,\varepsilon)+v) \qquad \mu = 1,\ldots,n$$

$$C(x,v,\varepsilon) = c(x,u^N(x,\varepsilon)+v) - c(x,u^N(x,\varepsilon)) + $$

$$+ \sum_{\mu=1}^{n}\left[b_\mu(x,u^N(x,\varepsilon)+v) - b_\mu(x,u^N(x,\varepsilon))\right]\frac{\partial u^N(x,\varepsilon)}{\partial x_\mu}$$

$$D(x,\varepsilon) = -A_\varepsilon u^N(x,\varepsilon)$$

and $[D(x,\varepsilon)]_0 \leq K\varepsilon^{N+1}$. Let (ϱ,φ) be local coordinates in a neighbourhood of $\Gamma_+(u_0)$ such that $\varrho(x) = \mathrm{dist}(x,\partial\Omega)$ and $0 < \varrho < \varrho_0$ corresponds to a strip parallel to $\partial\Omega$. (4.23) transforms into

$$\varepsilon \sum_{\kappa+|\beta|\leq 2} a_{\kappa\beta}(\varrho,\varphi) D_\varrho^\kappa D_\varphi^\beta v + B_0(\varrho,\varphi,v,\varepsilon) D_\varrho v + \sum_{\mu=1}^{n-1} \tilde{B}_\mu(\varrho,\varphi,v,\varepsilon) D_{\varphi_\mu} v +$$

$$+ \tilde{C}(\varrho,\varphi,\varepsilon,v) = \tilde{D}(\varrho,\varphi,\varepsilon)$$

$$v\big|_{\varrho=0} = -u^N\big|_{\partial\Omega} .$$

For the local corrector $(\zeta = \frac{\varrho}{\varepsilon})$

$$v^N(\zeta,\varphi,\varepsilon) = \sum_{\mu=0}^{N} \varepsilon^\mu v_\mu(\zeta,\varphi)$$

from this we obtain problems of the form

$$D_\zeta^2 v_0 + b^*(\varphi,v_0) D_\zeta v_0 = 0 \qquad (4.24)$$

$$v_0(0,\varphi) = -u_0\big|_{\varrho=0}, \qquad \lim_{\zeta\to\infty} v_0 = 0$$

$$D_\zeta^2 v_k + b^*(\varphi,v_0) D_\zeta v_k + \frac{\partial b^*}{\partial v}(\varphi,v_0) D_\zeta v_0 v_k = H_k(\zeta,\varphi,v_0,\dots,v_{k-1}) \tag{4.25}$$
$$v_k(0,\varphi) = -u_k\big|_{\varrho=0}, \qquad \lim_{\zeta\to\infty} v_k = 0, \quad k = 1,\dots,N$$

with $b^*(\varphi,v) = B_0(0,\varphi,v,0)/a_{20}(0,\varphi)$. Because $\partial\Omega = \Gamma_+(u_0)$ and $b^*(\varphi,0) = \left(\sum_{\mu=1}^n b_\mu(x,u_0(x)) \frac{\partial \varrho}{\partial x_\mu}\big|_{\varrho=0} \right) / a_{20}(0,\varphi)$ it holds $b^*(\varphi,0) > 0$. Additionally, we assume

(H 14) Let

$$q(\varphi,s) = \int_0^s b^*(\varphi,t)\,dt \neq 0$$

for all φ, s with $|s| \in (0, |u_0|_{\varrho=0}|]$, sgn s = - sgn($u_0|_{\varrho=0}$).

If $u_0|_{\varrho=0} = 0$ for $\varphi = \varphi_0$, $v_0(\zeta,\varphi_0) \equiv 0$ is the unique solution of (4.24). In the case of $u_0|_{\varrho=0} \neq 0$ we can express the solution of (4.24) in the implicit form

$$\zeta = \int_{v_0}^{-u_0|_{\varrho=0}} \frac{1}{q(\varphi,s)}\,ds\,. \tag{4.26}$$

From (4.26) we obtain the following asymptotic behaviour of v_0.
For all sufficiently small positive γ there are positive constants K_1 and K_2 independent of ζ such that

$$K_1 \exp\left(-[b^*(\varphi,0)+\gamma]\zeta\right) \le |D_\zeta^k D_\varphi^\beta v_0| \le K_2 \exp\left(-[b^*(\varphi,0)-\gamma]\zeta\right).$$

The solutions of the linear problems (4.25) can be determined as usual in the case $v_0(\zeta,\varphi_0) \equiv 0$. According to A 1.1 they are of classical boundary layer type. Now, let $u_0|_{\varrho=0} \neq 0$. Then we have
$E(\zeta,\varphi) = D_\zeta v_0(\zeta,\varphi) \neq 0$ and the solutions of problems (4.25) can be expressed in the form

$$v_k(\zeta,\varphi) = -\left(\frac{u_k|_{\varrho=0}}{E(0,\varphi)} + \int_0^\zeta \frac{1}{E(t,\varphi)} \int_t^\infty H_k(s,\varphi,v_0(s,\varphi),\dots,v_{k-1}(s,\varphi))\,ds\,dt \right) E(\zeta,\varphi)\,.$$

Because H_k is a linear combination of the v_i ($i \le k-1$) and their derivatives with coefficients depending on ζ, φ, v_0, by induction we obtain that $v_1, v_2, \dots, v_N$ are classical boundary layer functions.
Now we set

$$\phi_\varepsilon^N(x) = u^N(x,\varepsilon) + \psi(x)\, v^N(\zeta,\varphi,\varepsilon)$$

with a cut-off function $\psi \in C^\infty(\Omega)$ for which $0 \leq \psi(x) \leq 1$ and

$$\psi(x) = \begin{cases} 1 & \text{for } 0 \leq \rho(x) \leq \frac{\rho_0}{3} \\ 0 & \text{for } \rho(x) \geq \frac{2}{3}\rho_0 . \end{cases}$$

<u>Theorem 4.23.</u> Let (H 12), (H 13) and (H 14) be fulfilled. Then there are constants K_1 and K_2 independent of ε such that there exists a solution u of (4.21) uniquely determined in

$$B(\phi_\varepsilon^N, K_1\varepsilon) = \{ u_\varepsilon \in C^{2+\alpha}(\bar{\Omega}) \mid |u_\varepsilon - \phi_\varepsilon^N|_{2+\alpha} \leq K_1 \varepsilon \} .$$

ϕ_ε^N is a uniform asymptotic solution with

$$|u_\varepsilon - \phi_\varepsilon^N|_{2+\alpha} \leq K_2 \varepsilon^{N+1} . \qquad \blacksquare$$

<u>Proof.</u> For $z_\varepsilon = u_\varepsilon - \phi_\varepsilon^{N+1}$ we obtain the equivalent problem

$$A_\varepsilon(z_\varepsilon + \phi_\varepsilon^{N+1}) - A_\varepsilon \phi_\varepsilon^{N+1} = L_\varepsilon z_\varepsilon + R_\varepsilon z_\varepsilon = F_\varepsilon$$

with $[F_\varepsilon]_0 \leq K\varepsilon^{N+1}$, the linear operator

$$\tilde{L}_\varepsilon z \equiv \varepsilon L z + \sum_{\mu=1}^{n} b_\mu(x,\phi_\varepsilon^{N+1}) \frac{\partial z}{\partial x_\mu} + \Big[\sum_{\mu=1}^{n} \frac{\partial b_\mu}{\partial u}(x,\phi_\varepsilon^{N+1}) \frac{\partial \phi_\varepsilon^{N+1}}{\partial x_\mu} + \frac{\partial c}{\partial u}(x,\phi_\varepsilon^{N+1}) \Big] z$$

and the nonlinear operator

$$R_\varepsilon z \equiv \sum_{\mu=1}^{n} \big[b_\mu(x,\phi_\varepsilon^{N+1}+z) - b_\mu(x,\phi_\varepsilon^{N+1}) \big] \frac{\partial z}{\partial x_\mu} + \sum_{\mu=1}^{n} \big[b_\mu(x,\phi_\varepsilon^{N+1}+z) - b_\mu(x,\phi_\varepsilon^{N+1})$$
$$- \frac{\partial b_\mu}{\partial u}(x,\phi_\varepsilon^{N+1}) z \big] \frac{\partial \phi_\varepsilon^{N+1}}{\partial x_\mu} + c(x,\phi_\varepsilon^{N+1}+z) - c(x,\phi_\varepsilon^{N+1}) - \frac{\partial c}{\partial u}(x,\phi_\varepsilon^{N+1}) z .$$

The verification of the bounded Lipschitz continuity of R_ε can be shown in the same way as in the proof of Theorem 4.22. In difference to this proof it has to be taken into consideration that the term and its derivatives which must be estimated are not uniformly bounded with respect to ε. We have the estimate

$$|R_\varepsilon z_1 - R_\varepsilon z_2|_\alpha \leq m_\varepsilon(\sigma) |z_1 - z_2|_{2+\alpha} \qquad \text{with } m_\varepsilon(\sigma) = \frac{K\sigma}{\varepsilon}$$

for all σ with $0 \leq \sigma \leq \sigma^*$ and for all z_1, z_2 with $|z_1|_{2+\alpha} \leq \sigma$, $|z_2|_{2+\alpha} \leq \sigma$ where the norm in $C^\alpha(\bar{\Omega})$ is again defined by $|u|_\alpha = [u]_0 + \varepsilon^{1+\alpha}[u]_\alpha$.
We now show that $\tilde{L}_\varepsilon : C_0^{2+\alpha} \longrightarrow C^\alpha$ is a bijection and has a continuous

inverse $\tilde{L}_\varepsilon^{-1}$. The application of Theorem 2.14 first, yields

$$|z|_{2+\alpha} \leq K\left(\varepsilon^{1+\alpha} \|\tilde{L}_\varepsilon z\|_\alpha + [z]_0\right).$$

For estimating $[z]_0$ we note that

$$\left[\sum_{\mu=1}^{n} \frac{\partial b_\mu}{\partial u}(x, \phi_\varepsilon^{N+1}) \frac{\partial \phi_\varepsilon^{N+1}}{\partial x_\mu} + \frac{\partial c}{\partial u}(x, \phi_\varepsilon^{N+1})\right] \geq \frac{c_0}{2} > 0$$

for $\varrho(x) \geq \delta > 0$ and sufficiently small ε, however, in the case of $0 \leq \varrho(x) \leq \delta$ the sign of this expression can be arbitrary. Therefore the maximum principle cannot be used immediately. Consequently, we shall apply the maximum principle to a boundary value problem for $Z = \frac{z}{w_\varepsilon}$ with a suitable w_ε. Then it holds

$$L_\varepsilon^* Z \equiv \sum_{|\alpha| \leq 2} a_\alpha^*(x,\varepsilon) D^\alpha Z = \tilde{L}_\varepsilon z \qquad in\ \Omega$$
$$Z = 0 \qquad on\ \partial\Omega$$

with $a_0(x,\varepsilon) = \tilde{L}_\varepsilon w_\varepsilon$ and the main part of L_ε^* is equal to the product of w_ε with the main part of $\tilde{L}_\varepsilon$. Constructing w_ε in such a way that

$$\tilde{L}_\varepsilon w_\varepsilon \geq 1 \qquad in\ \Omega \tag{4.27}$$
$$0 < m_1 \leq w_\varepsilon \leq m_2 \qquad in\ \bar{\Omega}$$

we have $[\tilde{L}_\varepsilon z]_0$ as a barrier function for Z and it holds

$$[z]_0 \leq m_2 [Z]_0 \leq m_2 [\tilde{L}_\varepsilon z]_0 . \tag{4.28}$$

Now we shall construct a function w_ε with the properties (4.27). For this purpose let us suppose w_ε in the form

$$w_\varepsilon(x) = K_0 + \psi(x)[w_0(\zeta,\varphi) + \varepsilon w_1(\zeta,\varphi)].$$

For $\varrho(x) \geq \frac{2}{3}\varrho_0$ it holds $w_\varepsilon(x) = K_0$ and we have

$$|\tilde{L}_\varepsilon w_\varepsilon - c^*(x) K_0| \leq K\varepsilon .$$

Choosing K_0 sufficiently large, according to (H 12) it follows that

$$\tilde{L}_\varepsilon w_\varepsilon \geq c_0 K_0 - \varepsilon K \geq 1 \qquad in\ \{x \mid \varrho(x) \geq \tfrac{2}{3}\varrho_0\} .$$

In the domain $\{x \mid 0 \leq \varrho(x) \leq \frac{\varrho_0}{3}\}$, in local coordinates we obtain

$$\tilde{L}_\varepsilon w_\varepsilon \geq \frac{a_{20}(0,\varphi)}{\varepsilon}\left[D_\zeta^2 w_0 + b^*(\varphi,v_0) D_\zeta w_0 + \frac{\partial b^*}{\partial v}(\varphi,v_0) D_\zeta v_0 (w_0 + K_0)\right] +$$

$$+ a_{20}(0,\varphi)\left[D_\zeta^2 w_1 + b^*(\varphi,v_0) D_\zeta w_1 + \frac{\partial b^*}{\partial v}(\varphi,v_0) D_\zeta v_0 w_1 + \tilde{H}\right] + c_0 K_0 - \varepsilon K$$

provided that w_0 and w_1 are functions of classical boundary layer type. Let w_0 and w_1 be solutions of the problems

$$D_\zeta^2 w_0 + b^*(\varphi,v_0) D_\zeta w_0 + \frac{\partial b^*}{\partial v}(\varphi,v_0) D_\zeta v_0 w_0 = -K^* \exp\left(-[b^*(0,\varphi) - \gamma]\zeta\right)$$

$$w_0(0,\varphi) = 1, \qquad \lim_{\zeta\to\infty} w_0(\zeta,\varphi) = 0, \qquad \gamma \text{ arbitrary with } 0<\gamma<b^*(0,\varphi)$$

$$D_\zeta^2 w_1 + b^*(\varphi,v_0) D_\zeta w_1 + \frac{\partial b^*}{\partial v}(\varphi,v_0) D_\zeta v_0 w_1 = -\tilde{H}$$

$$w_1(0,\varphi) = 0, \qquad \lim_{\zeta\to\infty} w_1(\zeta,\varphi) = 0 .$$

We mention that the left-hand sides of the differential equations for w_0 and w_1 correspond to the left-hand sides of the differential equations (4.25). So w_0 is of classical boundary layer type and non-negative because of the nonpositivity of the right-hand side of the differential equation. Since an estimate of the form $|\tilde{H}| \leq K \exp(-\gamma^*\zeta)$ is valid for $\tilde{H}$, we have that also w_1 is of classical boundary layer type. Consequently, because of $a_{20}(0,\varphi) < 0$ and

$$\frac{\partial b^*}{\partial v}(\varphi,v_0) D_\zeta v_0 K_0 \leq K^* \exp\left(-[b^*(0,\varphi) - \gamma]\zeta\right)$$

we obtain

$$\tilde{L}_\varepsilon w_\varepsilon \geq c_0 K_0 - \varepsilon K \geq 1 \qquad \text{in } \{x \mid 0 \leq \varrho(x) \leq \tfrac{\varrho_0}{3}\}$$

for sufficiently small ε. In the remaining domain $\{x \mid \frac{\varrho_0}{3} \leq \varrho(x) \leq \frac{2}{3}\varrho_0\}$ we take into consideration that w_0, w_1 and v^N are transcendentally small. According to the construction of w_ε it now hold (4.27) and (4.28) with constants m_1 and m_2 independent of ε. This proves that $\tilde{L}_\varepsilon : C_0^{2+\alpha} \to C^\alpha$ is a bijection and has a continuous inverse $\tilde{L}_\varepsilon^{-1}$ with

$$|z|_{2+\alpha} \leq \frac{1}{l} |\tilde{L}_\varepsilon z|_\alpha .$$

To apply Theorem 2.13 we set $\sigma = \frac{l\varepsilon}{2K}$ such that for sufficiently small ε $|F_\varepsilon|_\alpha \le \frac{\sigma_0 l}{2}$ $(N \ge 1)$. Let $K_1 = \frac{1}{2K}$ and $K_2 = \frac{2}{l}$. Then there exists a solution u_ε of (4.21) uniquely determined in $B(\phi_\varepsilon^{N+1}, K_1\varepsilon)$ and it holds

$$|u_\varepsilon - \phi_\varepsilon^{N+1}|_{2+\alpha} \le K_2\, \varepsilon^{N+1} .$$

Because $|\phi_\varepsilon^{N+1} - \phi_\varepsilon^{N}|_{2+\alpha} \le K\, \varepsilon^{N+1}$ we also have

$$|u_\varepsilon - \phi_\varepsilon^{N}|_{2+\alpha} \le K\, \varepsilon^{N+1} .$$

Furthermore, $B(\phi_\varepsilon^{N}, K_1^*\varepsilon)$ is contained in $B(\phi_\varepsilon^{N+1}, K_1\varepsilon)$ if K_1^* is suitably chosen. This concludes the proof of Theorem 4.23 . ∎

__Remark.__ If we choose the function w_0 appearing in the proof as solution of

$$D_\zeta^2 w_0 + b^*(\varphi, V_0) D_\zeta w_0 + \frac{\partial b^*}{\partial v}(\varphi, V_0) D_\zeta V_0 (w_0 + K_0) = 0$$

$$w_0(0, \varphi) = \widetilde{K}, \qquad \lim_{\zeta \to \infty} w_0(\zeta, \varphi) = 0 ,$$

then also w_0 and w_1 are of classical boundary layer type, however the assertion of VAN HARTEN (1978), that $w_0 > 0$ for sufficiently large $\widetilde{K}$, is not valid. ∎

Comments to Chapter 4

4.1. The problem to determine the subdomains in which the solution of the formal global problem (frequently also called "reduced" problem) is a uniform asymptotic solution has little been considered in the literature.

The first treatise on this subject for elliptic equations of second order has been given by LUBE (1980), where, partially, also boundary conditions of second and third kind had been investigated.

HOWES (1979a,b), (1980) has stated similar results as that given in the Theorems 4.1, 4.2, in the case of nonlinear equations of the type considered in 4.6; however, it is always assumed that $b \cdot \nabla F \geq 0$ on the boundary.

An estimation technique suggested by GÖHDE (1981) makes it possible, perhaps, to extend the results to equations of higher order.

A global statement related to Theorem 4.4 can already be found in ARONSON (1956) for parabolic equations. Theorem 4.5 for n = 1 has been proved by BOBISUD (1968) with the help of the estimation technique used in Section 4.3.2. For arbitrary space dimension FELGENHAUER (1980) has applied a weak maximum principle.

VENTCEL'/FREJDLIN (1979) did not give uniform estimates, however they studied pointwise convergence by means of impressive probabilistic methods.

4.2. In the classical works on singular perturbed elliptic boundary value problems it was usual to start from

$$\varepsilon L_2 u_\varepsilon + \frac{\partial u_\varepsilon}{\partial x_2} = f .$$

Then, the behaviour of the characteristics is very simple, namely the characteristics are the straight lines x_1 = const. Nevertheless, serious difficulties arise, since for regular boundary Γ_- is not closed in Γ and for nonregular boundary (for instance, if Ω is a parallelepiped) we are confronted with problems in the corners.

Our approach is motivated by the fact that the simplest case is not characterized by simplest behaviour of the characteristics, but that the behaviour of the characteristics is most uniformly possible in the neighbourhood of the boundary and furthermore the solution of the global formal problem is regular. Then a "very simple" problem, for instance, is

$$\varepsilon L_2 u_\varepsilon \pm \sum_{i=1}^{n} x_i \frac{\partial u_\varepsilon}{\partial x_i} + c u_\varepsilon = f$$

with positive c, where Ω denotes the unit circle (cf. 4.2.1 and 4.2.2).

The results of 4.2 are also important for such problems, in which Γ is only piecewise regular. Naturally, additional correctors in the neighbourhood of the corners have to be constructed. Results concerning this subject can be found in BUTUSOV (1975), (1977), (1979), IL'IN/LELIKOVA (1975), LELIKOVA (1978), (1979).

4.3. BESJES (1975a) was the first who investigated a singularly perturbed problem for an elliptic equation of arbitrary order, namely

$$\varepsilon L_2 u_\varepsilon + \frac{\partial u_\varepsilon}{\partial x_n} = f .$$

Thereby it was necessary to exclude neighbourhoods of Γ_0 (Γ is regular). DE RUIJTER (1979) has applied the results of BESJES to problems of oceanography. His considerations are, partially, only formal.

A special parabolic boundary value problem of arbitrary order has been considered by ISAKOVA 1966.

4.4. GRASMAN (1971) has dealt with the problem of tangent characteristics in the case of elliptic equations of second order. Thereby the characteristics are tangential from outside and the order of contact is arbitrary. An essential simplification was independently given by VAN HARTEN (1976) and LELIKOVA (1976) (the order of contact is one). The representation in 4.4.1 is based on the latter work.

IL'IN, GOR'KOV/LELIKOVA (1975) have studied problems in which the characteristics are tangential from the interior and a "free boundary layer" appears, namely

$$\varepsilon \Delta u_\varepsilon - y \frac{\partial u_\varepsilon}{\partial x} - \frac{\partial u_\varepsilon}{\partial y} = f$$

in a subdomain of $0 < x < 2\delta$; the characteristic $x = \frac{y^2}{2}$ is tangential to the domain from the interior at (0,0).

SCHAIGARDANOV (1979) considered a problem of order four, where the characteristics are tangential from the interior, and applied the technique of BESJES for the proof.

4.5. For estimating the principal eigenvalue of

$$L_\varepsilon \equiv \varepsilon \sum_{i,j=1}^{n} a_{ij} \frac{\partial^2}{\partial x_i \partial x_j} + \sum_{i=1}^{n} b_i \frac{\partial}{\partial x_i}$$

in the space of Hölder-continuous functions mainly probalistic methods have been used in the case of the first boundary value problem. FRIEDMAN (1973), VENTCEL (1975) have proved that $\hat{\tau}_\varepsilon \geq \frac{K}{\varepsilon}$ provided that all characteristics leave $\overline{\Omega}$ in finite time. The remaining estimates, given above for the cases that turning points do exist, can be found in DEVINATZ/FRIEDMAN (1974),(1978a).

Precise estimates in the case that $\Gamma = \Gamma_+$ (then $\hat{\tau}_\varepsilon$ converges exponentially to zero) were obtained by VENTCEL/FREIDLIN (1979) (Theorem 7.4), KAMIN (1979b).
The given examples with

$$L_0 \equiv \pm\left(x_1 \frac{\partial}{\partial x_1} + b_2 x_2 \frac{\partial}{\partial x_2}\right)$$

in the unit circle come from DE GROEN (1976). There also statements for

$$L_0 \equiv \left[(x_1 - l x_2)\frac{\partial}{\partial x_1} + (x_2 + l x_1)\frac{\partial}{\partial x_2}\right]$$

and finally for

$$L_0 \equiv x_1 \frac{\partial}{\partial x_1} - l x_2 \frac{\partial}{\partial x_2}$$

in the unit square can be found.
The point (0,0) is a saddle point in the last case; more general estimates for such cases are given in KIFER (1980).
Estimates of the principal eigenvalue for the second and third boundary value problem are the main objective of the work of HINTZE (1981).
The treatment of singularly perturbed problems for which the solution of the global formal problem is not uniquely determined, has started with the beginning of the 70ies years (DORR (1970), ACKERBERG/O'MALLEY (1970), COOK/ECKHAUS (1973)) with one-dimensional problems. In the case of $n = 1$ the subject is almost closed for simple turning points. (DE GROEN (1976), OLVER (1978)). MATKOWSKY/SCHUSS (1976), (1977), GRASMAN/MATKOWSKY (1977) considered

$$\varepsilon \sum_{i,j=1}^{n} a_{ij} \frac{\partial^2 u_\varepsilon}{\partial x_i \partial x_j} + \sum_{i=1}^{n} b_i \frac{\partial u_\varepsilon}{\partial x_i} = 0\,, \qquad u_\varepsilon\big|_\Gamma = g \tag{4.18}$$

on condition that $\Gamma = \Gamma_+$ and that exactly one turning point exists. They conjectured that $\lim_{\varepsilon \to 0} u_\varepsilon = K$ in the interior of Ω and tried to determine the constant K, for instance, with the help of variational formulations.
Exact proofs were given by KAMIN (1978), (1979a). They are stated in 4.5.1.1. The idea used in 4.5.1.1 to determine the constant K was for the first time to be found in DEVINATZ/FRIEDMAN (1978b). In the work of KAMIN the constant K was determined in a more complicated way without the assumption (H 7).
GRASMAN (1979) has investigated problems of the type (4.18) containing a parameter vector. In other works, which are frequently only formal, it was tried to allow more than one inner turning point

(SCHUSS/MATKOWSKY (1979), MANGEL (1979), (1980), VENTCEL/FREIDLIN (1979), WILLIAMS (1981))
A special problem with closed characteristics has already been studied by CHASMINSKIJ (1963), problems as in 4.5.1.4 have been considered by DE GROEN (1974/75).
Special problems with singular lines were also investigated by BARTON (1976) (special case of 4.5.2.1), DE JAGER (1972) (special case of Theorem 4.19c) and GRASMAN (1977). GRASMAN's assertion that for k = 1 it holds s = 1 in Theorem 4.21 is not true.

4.6. Up to now three methods for investigating nonlinear singular perturbed elliptic problems are known: Using the implicit function theorem (FIFE (1973)), applying the fixed point theorem of Banach (VAN HARTEN (1976), (1978)) and estimating by means of sub-and supersolutions (HOWES (1979a,b), (1980)).
Special problems of the type considered here have first been investigated by VAN HARTEN (1976), (1978). There also a problem with characteristics which are tangential from outside has been studies.
HOWES (1979a,b), (1980) and HABETS (1980) have determined sub-domains in which the solution of the global problem is a uniform asymptotic solution and obtain results similar to those in section 4.1.

Appendix: Boundary layer problems

We shall deal with problems of the form

$$L v_0 = 0, \qquad L v_\mu = F_\mu$$

where the $F_\mu = F_\mu(v_0, \ldots, v_{\mu-1})$ are multilinear differential operators the coefficients of which are polynomials in ζ. For the boundary layer variable ζ it holds $\zeta \in [0, \infty)$. We are interested in classical solutions which, with respect to ζ, fulfil conditions at $\zeta = 0$, for instance

$$v_\mu^{(s)}\Big|_{\zeta=0} = g_{\mu s}, \qquad s = 0, 1, \ldots, s^*$$

and decay in some sense for $\zeta \to \infty$. v_μ are said to be of classical boundary layer type if

$$\lim_{\zeta\to\infty} \zeta^l v_\mu = 0$$

for arbitrary l and analogous statements are valid for some derivatives of v_μ.

A 1 Ordinary boundary layers

1.1. Constant coefficients (1)

Let

$$L v \equiv (-1)^m \frac{d^{2m} v}{d\zeta^{2m}} + b \frac{dv}{d\zeta}$$

and let b be a real constant which is nonzero. The corresponding characteristic equation is

$$(-1)^m \varkappa^{2m} + b\varkappa = 0 .$$

It now holds that

$$\varkappa_k = \sqrt[2m-1]{|b|}\ \exp\left(\frac{2k}{2m-1}\pi i\right) \begin{cases} \exp\left(\frac{m-1}{2m-1}\pi i\right), & b > 0 \\ & k = 0, \ldots, 2m-1 \\ \exp\left(\frac{m}{2m-1}\pi i\right), & b < 0 \end{cases},$$

that means in the case of $b < 0$ we have $\operatorname{Re} \varkappa_k < 0$ if and only if $k = 0, 1, \ldots, m-1$ and otherwise, in the case of $b > 0$, we have $\operatorname{Re} \varkappa_k < 0$ if and only if $k = 1, \ldots, m-1$.

Consequently, solutions of classical boundary layer type do exist if for $b < 0$ m additional conditions at $\zeta = 0$ and for $b > 0$ m-1 additional conditions are required.

1.2. Constant coefficients (2)

Let

$$Lv \equiv (-1)^m \frac{d^{2m}v}{d\zeta^{2m}} + cv$$

where c has a positive real part. The corresponding characteristic equation is

$$(-1)^m \varkappa^{2m} + c = 0$$

It holds

$$\varkappa_k = \sqrt[2m]{|c|}\ \exp\left(\frac{2k}{2m}\pi i + \frac{m+1+\varphi}{2m}\pi i\right),\quad k = 0,1,\dots,2m-1$$

with $\quad -\frac{\pi}{2} < \arg c = \varphi < \frac{\pi}{2}$.

Consequently we have Re $\varkappa_k < 0$ if and only if $k = 0,1,\dots,m-1$. Requiring m additional conditions at $\zeta = 0$, solutions of classical boundary layer type do exist.

1.3. Variable coefficients (1)

Let specially

$$Lv \equiv -\frac{d^2v}{d\zeta^2} + b\zeta\frac{dv}{d\zeta} + cv$$

where b and c are constants. The solution of

$$Lv = 0, \qquad v\big|_{\zeta=0} = g$$

can be expressed then, for instance, by parabolic cylinder functions, it holds

$$v = \sqrt{\frac{2}{\pi}}\ g\exp\left(\frac{b}{4}\zeta^2\right)\cdot\begin{cases}\dfrac{\Gamma\left(\frac{2-\frac{c}{b}}{2}\right)}{2^{\frac{c}{b}-1}} D_{\frac{c}{b}-1}\left(\sqrt{-b}\,\zeta\right), & b<0\\[2ex] \dfrac{\Gamma\left(\frac{1+\frac{c}{b}}{2}\right)}{2^{-\frac{c}{b}}} D_{-\frac{c}{b}}\left(\sqrt{b}\,\zeta\right), & b>0\end{cases}$$

where Γ denotes the Gamma function. The behaviour for $\zeta \to \infty$ is characterized by

$$v \sim \begin{cases} K_1 \zeta^{\frac{c}{b}-1} \exp\left(\frac{b}{2}\zeta^2\right), & b < 0 \\ K_2 \zeta^{-\frac{c}{b}}, & b > 0 . \end{cases}$$

Only in the case $b < 0$ we obtain a solution of classical boundary layer type, for $b > 0$ we have a decaying for $\zeta \to \infty$ at least as $\zeta^{-\frac{c}{b}}$ for positive c.

Considering, more generally, that

$$Lv \equiv (-1)^m \frac{d^{2m}v}{d\zeta^{2m}} + b\zeta \frac{dv}{d\zeta} + cv$$

and the boundary value problem

$$Lv = 0 , \qquad v^{(s)}\Big|_{\zeta=0} = g_s , \quad s = 0, 1, \dots, m-1$$

we can obtain the asymptotic behaviour of solutions from the corresponding Newton polygon. From this solutions with

$$v \sim \exp(\varkappa \zeta^{\frac{2m}{2m-1}})$$

do exist, where $\varkappa$ satisfies

$$(-1)^m \left(\frac{2m}{2m-1}\right)^{2m} \varkappa^{2m} + \frac{2m}{2m-1} b \varkappa = 0 .$$

Exactly as in A 1.1 there exist m $\varkappa_k$ with negative real part for $b < 0$ and m-1 $\varkappa_k$ with negative real part for $b > 0$. However, as for $m = 1$, for $b > 0$, in addition, there exists one solution behaving like $\zeta^{-\frac{c}{b}}$. Consequently, the above boundary layer problem has a unique solution for arbitrary m, which decays for $\zeta \to \infty$ but the effective calculation is difficult because we have to solve, for instance, the so-called central connecting problem (see for instance ROOS (1979), SIBUYA (1975)). The treatment of inhomogenous problems $Lv_\mu = F_\mu$ for equations with an order greater than two is an open problem.

1.4. Variable coefficients (2)

Specially we consider

$$Lv \equiv -\frac{d^2v}{d\zeta^2} + b\zeta^l \frac{dv}{d\zeta} = 0$$
$$v\Big|_{\zeta=0} = g$$

where b is a constant and l is a positive integer.
Only for $b < 0$ there exists a decaying solution, namely

$$v = g \frac{\int_{\infty}^{\zeta} \exp\left(\frac{b}{l+1} t^{l+1}\right) dt}{\int_{\infty}^{0} \exp\left(\frac{b}{l+1} t^{l+1}\right) dt} \quad \text{with } v \sim K \zeta^{-l} \exp\left(\frac{b}{l+1} \zeta^{l+1}\right) \text{ for } \zeta \to \infty.$$

The relation

$$\int_0^{\infty} \exp(-x^{\mu})\, dx = \frac{1}{\mu} \Gamma\left(\frac{1}{\mu}\right)$$

yields

$$v = \frac{g}{A} \int_{\zeta}^{\infty} \exp\left(\frac{b}{l+1} t^{l+1}\right) dt$$

with

$$A = \left(-\frac{b}{l+1}\right)^{-\frac{1}{l+1}} \frac{1}{l+1} \Gamma\left(\frac{1}{l+1}\right) .$$

A 2 Parabolic boundary layers (I)

2.1. The simplest case of an equation of second order

A parabolic boundary layer problem has first been investigated by ECKHAUS (1966), namely

$$L v_0 \equiv -\frac{\partial^2 v_0}{\partial \zeta^2} + \frac{\partial v_0}{\partial t} = 0$$

$$v_0 \Big|_{t=0} = 0$$

$$v_0 \Big|_{\zeta=0} = g(t) .$$

The solution v_0 can explicitly be expressed in the form

$$v_0(\zeta, t) = \sqrt{\frac{2}{\pi}} \int_{\frac{\zeta}{\sqrt{2t}}}^{\infty} e^{-\frac{1}{2}\alpha^2} g\left(t - \frac{\zeta^2}{2\alpha^2}\right) d\alpha .$$

Both v_0 and its derivatives are classical boundary layer functions. On the compatibility condition that $g(0) = 0$,

$v_0, \frac{\partial v_0}{\partial t}, \frac{\partial v_0}{\partial \zeta}, \frac{\partial^2 v_0}{\partial \zeta^2}$ are uniformly bounded.

However, it holds

$$\frac{\partial^2 v_0}{\partial t^2} = \sqrt{\frac{2}{\pi}}\left[\frac{\zeta}{(2t)^{3/2}} g'(0) \exp\left(-\frac{1}{2}\frac{\zeta^2}{2t}\right) + \int_{\frac{\zeta}{\sqrt{2t}}}^{\infty} e^{-\frac{1}{2}\alpha^2} g''\left(t - \frac{\zeta^2}{2\alpha^2}\right) d\alpha\right],$$

that means $\frac{\partial^2 v_0}{\partial t^2}$ is uniformly bounded if and only if $g'(0) = 0$. The analogous is true for $\frac{\partial^2 v_0}{\partial t \partial \zeta}$. This affects the solutions of the inhomogenous problems

$$L v_\mu = F_\mu .$$

However, supposing that $g^{(k)}(0) = 0$ for arbitrary k, we can prove that all v_μ and their derivatives are uniformly bounded and of classical boundary layer type.

2.2. The simplest case of an equation of arbitrary order

The just stated results can be generalized on 2m-parabolic boundary layer equations. Let

$$L v_0 \equiv (-1)^m \frac{\partial^{2m} v_0}{\partial \zeta^{2m}} + \frac{\partial v_0}{\partial t} = 0$$

$$v_0\Big|_{t=0} = 0$$

$$\frac{\partial^s v_0}{\partial \zeta^s}\Big|_{\zeta=0} = g_s(t), \qquad s = 0, \ldots, m-1 .$$

Then v_0 can be expressed in the form

$$v_0 = \sum_{s=0}^{m-1} \int_0^t \mu_s(\tau) \frac{\partial^s \Gamma(\zeta, t, \xi, \tau)}{\partial \xi^s}\Big|_{\xi=0} d\tau .$$

Thereby $\Gamma(\zeta, t, \xi, \tau)$ denotes the fundamental solution which corresponds to L

$$\Gamma = \frac{1}{\pi}\int_0^\infty \cos \alpha(\zeta - \xi)\, e^{-\alpha^{2m}(t-\tau)} d\alpha ,$$

for m = 1 it especially holds

$$\Gamma = \frac{1}{2\sqrt{\pi}}(t-\tau)^{-\frac{1}{2}} \exp\left(-\frac{(\zeta-\xi)^2}{4(t-\tau)}\right) .$$

The $\mu_k(t)$ are uniquely determined and regular if the $g_s(t)$ satisfy corresponding compatibility conditions at $t = 0$. Γ satisfies the basic estimate (EIDELMANN (1956), MICHAILOV (1960))

$$\left| \frac{\partial^k \Gamma(\zeta,t,\xi,\tau)}{\partial \zeta^k} \right| \leq K|t-\tau|^{-\frac{1+k}{2m}} exp\left[-\beta\left(\frac{(\zeta-\xi)^{2m}}{t-\tau}\right)^{\frac{1}{2m-1}}\right], \beta>0.$$

From this ISAKOVA (1966) has concluded that v_0 and its derivatives are uniformly bounded classical boundary layer functions, especially it holds

$$\left|\frac{\partial^k v_0}{\partial \zeta^k}\right| \leq K t^{-\frac{k}{2m}} exp\left(-\beta' \left(\frac{\zeta^{2m}}{t}\right)^{\frac{1}{2m-1}}\right), \quad k=0,\dots,2m-1, \quad \beta'>0.$$

Now we can prove by induction that also for the v_μ estimates of this type are valid, For instance, v_1 satisfies

$$L v_1 = F_1(v_0)$$

$$v_1\Big|_{t=0} = 0$$

$$\frac{\partial^s v_1}{\partial \zeta^s}\Big|_{\zeta=0} = g_{1s}(t)$$

and can be expressed in the form

$$v_1 = v_1^* + \tilde{v}_1$$

where v_1^* satisfies the homogenous problem with the data of v_1 and $\tilde{v}_1$ satisfies the inhomogenous problem with homogenous data. Obviously, for v_1^* it holds the estimate obtained for v_0. $F_1(v_0)$ is a linear combination of $\frac{\partial^k v_0}{\partial \zeta^k}$ ($k = 0,1,\dots,2m-1$) the coefficients of which are polynomials in ζ. Consequently with $\beta'' < \beta'$ it holds that

$$|F_1(v_0)| \leq K t^{-\frac{i}{2m}} exp\left(-\beta''\left(\frac{\zeta^{2m}}{t}\right)^{\frac{1}{2m-1}}\right), \qquad i \leq 2m-1.$$

Now, $\tilde{v}_1$ can be expressed in the form

$$\tilde{v}_1 = \int_0^t \int_0^\infty G(\zeta,t,\xi,\tau)\left(F_1(v_0)\right)(\xi,\tau)\, d\xi\, d\tau$$

where Green's function $G(\zeta,t,\xi,\tau)$ satisfies the same estimate as $\Gamma(\zeta,\xi,t,\tau)$. Introducing the notation

$$G_k(\beta^*,\zeta,v) = v^{-\frac{k}{2m}} exp\left(-\beta^*\left(\frac{\zeta^{2m}}{v}\right)^{\frac{1}{2m-1}}\right)$$

we can estimate $\tilde{v}_1$ as follows

$$\left|\frac{\partial^k \tilde{v}_1}{\partial \zeta^k}\right| \le K \int_0^t \int_0^\infty \tau^{-\frac{i}{2m}} G_0(\beta'', \xi, \tau)\, G_{k+1}(\beta, \zeta-\xi, t-\tau)\, d\xi\, d\tau, \quad k=0,\dots,2m,\ i \le 2m-1.$$

Exactly this integral has been estimated by ISAKOVA (1966) and it has been obtained that

$$\left|\frac{\partial^k \tilde{v}_1}{\partial \zeta^k}\right| \le K t^{-\frac{k}{2m}} \exp\left(-\beta''\left(\frac{\zeta^{2m}}{t}\right)^{\frac{1}{2m-1}}\right), \qquad k=0,\dots,2m-1.$$

Consequently all v_ℓ and their derivatives with respect to ζ are classical boundary layer functions.
In the case of $m = 2$ explicit representations for v_0 have been investigated by DE RUIJTER (1979).

2.3. Periodicity in an additional independent variable

We consider

$$Lv \equiv (-1)^m \frac{\partial^{2m} v}{\partial \zeta^{2m}} + \frac{\partial v}{\partial t} + \frac{1}{\gamma(\varphi)} \frac{\partial v}{\partial \varphi} = F(\zeta, t, \varphi)$$

$$v\big|_{t=0} = 0$$

$$\frac{\partial^s v}{\partial \zeta^s}\bigg|_{\zeta=0} = g_s(t, \varphi), \quad s=0,\dots,m-1,$$

where $\gamma(\varphi)$ is periodic in φ. Let, without restriction of generality, $0 < \varphi < 2\pi$. We are interested in 2π-periodic solutions in φ. v can be expressed in the form

$$v = \sum_{k=-\infty}^{\infty} V_k(\zeta, t)\, \phi_k(\varphi),$$

the $\phi_k(\varphi)$ are periodic solutions of

$$\frac{d\phi}{d\varphi} = \mu \gamma(\varphi) \phi .$$

From this it follows

$$\phi_k(\varphi) = \exp\left(\mu_k \int_0^\varphi \gamma(\alpha)\, d\alpha\right), \qquad \mu_k = \frac{2\pi k i}{P}, \qquad p = \int_0^{2\pi} \gamma(\alpha)\, d\alpha .$$

In the special case $\gamma \equiv 1$ we have $\phi_k(\varphi) = \exp(ik\varphi)$. So the representation of v is a Fourier-series.

The $V_k(\zeta, t)$ satisfy

$$(-1)^m \frac{\partial^{2m} V_k}{\partial \zeta^{2m}} + \frac{\partial V_k}{\partial t} + \mu_k V_k = f_k(\zeta, t)$$

$$V_k\Big|_{t=0} = 0$$

$$\frac{\partial^s V_k}{\partial \zeta^s}\Big|_{\zeta=0} = g_{ks}(t)$$

where

$$f_k(\zeta, t) = \frac{1}{P}\int_0^{2\pi} \gamma(\varphi)\, F(\zeta, t; \varphi)\, \phi_{-k}(\varphi)\, d\varphi$$

$$g_{ks}(t) = \frac{1}{P}\int_0^{2\pi} \gamma(\varphi)\, g_s(t, \varphi)\, \phi_{-k}(\varphi)\, d\varphi .$$

Setting $V_k = \exp(-\mu_k t) \cdot \tilde{V}_k$ we can see that the $\tilde{V}_k$ satisfy the same problems investigated in section 2.2.

2.4. A more general equation of second order

Another generalisation of the boundary layer problem studied in A 2.1 is the boundary value problem

$$Lv \equiv -\frac{\partial^2 v}{\partial \zeta^2} + \frac{\partial v}{\partial t} + b(\varphi, t)\zeta \frac{\partial v}{\partial t} + q(\varphi, t)\frac{\partial v}{\partial \varphi} = 0$$

$$v\Big|_{t=0} = 0$$

$$v\Big|_{\zeta=0} = g(\varphi, t)$$

where b, q, g are periodic in φ.
Analogous to 2.1 we can give the solution in an explicit way.
Let $\tilde{\psi}(\varphi, \tau, t)$ be the solution of

$$\frac{d\psi}{dt} = g(\psi, t) , \qquad \psi(\tau) = \varphi$$

and let $s(\tau, t)$ be defined by

$$s(\tau, t, \varphi) = -\int_t^{\tau} \exp\Big(-2\int_t^{\beta} b\big(\tilde{\psi}(\varphi, \alpha, t), \alpha\big)\, d\alpha\Big)\, d\beta , \quad 0 \leq \tau \leq t .$$

Obviously, it holds

$$s > 0 \quad \text{for} \quad \tau < t \quad \text{and} \quad K_1|t-\tau| \leq |s(\tau,t,\varphi)| \leq K_2|t-\tau| .$$

Setting finally

$$\frac{\zeta^2}{4\alpha^2} = s(\tau,t,\varphi) \qquad \text{and} \quad \tau = s^*(\zeta,t,\alpha\,\varphi), \text{ respectively}$$

we obtain the representation

$$v(\zeta,t,\varphi) = \frac{2}{\sqrt{\pi}} \int\limits_{\frac{\zeta}{2\sqrt{s(\tau,t,\varphi)}}}^{\infty} e^{-\alpha^2} g\big(\tilde{\varphi}(\varphi,s^*,t), s^*\big)\, d\alpha \ .$$

For $b \equiv 0$ it holds $s = t - \tau$ and $s^* = t - \frac{\zeta^2}{4\alpha^2}$. This results in the representation of section 2.1. Because of the above given properties of $s(\tau,t,\varphi)$ we have the same properties for v as in the special case $b \equiv 0$, $q \equiv 0$.
For $b(\varphi,t) \equiv b$, $q \equiv 1$, for instance, it holds that

$$\tilde{\varphi} = \varphi + t - \tau$$

$$s = \frac{1}{2b}\left(e^{2b(t-\tau)} - 1\right)$$

$$s^* = t - \frac{1}{2b}\ln\left(1 + \frac{b}{2\alpha^2}\zeta\right) .$$

A 3 Parabolic boundary layers (II)

3.1. Periodic solutions of equations of second order in the simplest case

We are interested in a solution periodic in φ $(0 < \varphi < 2\pi)$ of

$$Lv \equiv -a(\varphi)\frac{\partial^2 v}{\partial \zeta^2} + q(\varphi)\frac{\partial v}{\partial \varphi} + cv = 0$$

$$v\Big|_{\zeta=0} = g(\varphi)$$

where q, g are periodic in φ, $q(\varphi) > 0$ and c is a positive constant. This problem only unessentially differs from that considered by DE GROEN (1974/75) where $q \equiv 1$.

Fourier-Sinus-transformation

$$v(\zeta,\varphi)=\frac{2}{\pi}\int_0^{\infty} w(z,\varphi)\sin z\zeta\, dz$$

produces the image equation

$$q(\varphi)\frac{dw}{d\varphi}+\left(c+a(\varphi)z^2\right)w=z\,a(\varphi)\,g(\varphi).$$

Using the abbreviations

$$A(z,\varphi)=\frac{c+a(\varphi)z^2}{q(\varphi)}, \qquad B(z,\varphi)=z\,\frac{a(\varphi)\,g(\varphi)}{q(\varphi)}$$

the image equation has the periodic solution

$$w(z,\varphi)=\frac{\exp\left(-\int_0^{2\pi}A(z,\alpha)\,d\alpha\right)}{1-\exp\left(-\int_0^{2\pi}A(z,\alpha)\,d\alpha\right)}\cdot\int_{\varphi}^{\varphi+2\pi}B(z,\alpha)\left[\exp\int_{\varphi}^{\alpha}A(z,\beta)\,d\beta\right]d\alpha .$$

Since w behaves only like the rational function

$$\frac{z}{c+a(\varphi)z^2}$$

for $z\to\infty$, $w(z,\varphi)$ is transformed by partial integration into

$$w(z,\varphi)=\frac{B(z,\varphi)}{A(z,\varphi)}+w^*(z,\varphi)=a(\varphi)\,g(\varphi)\,\frac{z}{c+a(\varphi)z^2}+w^*(z,\varphi).$$

Now w^* behaves already like

$$\frac{z}{[c+a(\varphi)z^2]^2}$$

and the analogous is true for the first and second derivatives of w^* with respect to φ. Transforming the corresponding decomposition of w back, this yields

$$v(\zeta,\varphi)=g(\varphi)\exp\left(-\zeta\sqrt{\frac{c}{a(\varphi)}}\right)+v^*(\zeta,\varphi)$$

where $v^*(\zeta,\varphi)$ is a classical boundary layer function, consequently, naturally also v.

3.2. Periodic solutions for an equation of arbitrary order

We are interested in a periodic solution ($0 < \varphi < 2\pi$) of

$$L v \equiv (-1)^m \frac{\partial^{2m} v}{\partial \zeta^{2m}} + \frac{1}{\gamma(\varphi)} \frac{\partial v}{\partial \varphi} + c v = 0$$

$$\left. \frac{\partial^s v}{\partial \zeta^s} \right|_{\zeta=0} = g_s(\varphi), \qquad s = 0, 1, \dots, m-1.$$

Because of the given boundary conditions the Fourier-Sinus-transformation is not suitable.

Let

$$v(\zeta, \varphi) = \sum_{k=-\infty}^{\infty} V_k(\zeta)\, \phi_k(\varphi).$$

Then, as in A 2.3, for the ϕ_k it holds

$$\phi_k(\varphi) = \exp\left(\mu_k \int_0^{\varphi} \gamma(\alpha)\, d\alpha\right), \qquad \mu_k = \frac{2\pi k i}{P}, \qquad P = \int_0^{2\pi} \gamma(\alpha)\, d\alpha$$

and the $V_k(\zeta)$ satisfy

$$\tilde{L} V_k \equiv (-1)^m \frac{d^{2m} V_k}{d \zeta^{2m}} + (c + \mu_k) V_k = 0$$

$$\left. \frac{d^s V_k}{d \zeta^s} \right|_{\zeta=0} = g_{ks}$$

with

$$g_{ks} = \frac{1}{P} \int_0^{2\pi} \gamma(\varphi)\, g_s(\varphi)\, \phi_{-k}(\varphi)\, d\varphi .$$

Let V_{ks} ($s = 0, \dots, m-1$) be the solution of

$$\tilde{L} V_{ks} = 0$$

$$\left. \frac{d^l V_{ks}}{d \zeta^l} \right|_{\zeta=0} = \delta_{ls} .$$

The V_{ks} and their derivatives decay exponentially for $\zeta \to \infty$.
For $k \to \infty$ $V_{ks}^{(l)}$ behaves like

$$\left[\sqrt{\lambda^2 + \frac{2\pi k}{P}} \right]^{\frac{l}{2m}} .$$

Then V_k can be expressed as

$$V_k = \sum_{s=0}^{m-1} g_{ks} V_{ks} .$$

The character of convergence of

$$\sum_{k=-\infty}^{\infty} V_k(\zeta)\, \phi_k(\varphi)$$

depends on the regularity of $g_s(\varphi)$. If the first l derivatives of $g_s(\varphi)$ are continuous, then the g_{ks} have the order of magnitudes $\frac{1}{k^l}$ for $k \to \infty$. Consequently, assuming that the $g_s(\varphi)$ are sufficiently regular, $v(\zeta, \varphi)$ is a classical boundary layer function.

Now we shall study the corresponding inhomogenous problems.

Let the just investigated v be v_o now. We consider

$$L v_1 = F_1(v_o)$$

$$\left.\frac{\partial^s v_1}{\partial \zeta^s}\right|_{\zeta=0} = 0, \qquad s = 0, 1, \dots, m-1 .$$

Let

$$v_1(\zeta, \varphi) = \sum_{k=-\infty}^{\infty} w_k(\zeta)\, \phi_k(\varphi) .$$

Then it holds

$$\tilde{L} w_k = F_{1,k}$$

$$\left.\frac{d^s w_k}{d \zeta^s}\right|_{\zeta=0} = 0 , \qquad s = 0, 1, \dots, m-1$$

with

$$F_{1,k} = \frac{1}{P} \int_0^{2\pi} \gamma(\varphi)\, F_1\, \phi_{-k}(\varphi)\, d\varphi .$$

If $F_1(v_o)$ does not depend on φ, from

$$v_o = \sum_{k=-\infty}^{\infty} \sum_{s=0}^{m-1} g_{ks} V_{ks}(\zeta)\, \phi_k(\varphi)$$

we immediately obtain

$$F_{1,k} = \sum_{s=0}^{m-1} g_{ks} P_s(\zeta) V_{ks}(\zeta) \qquad (*)$$

where $P_s(\zeta)$ denotes a polynomial in ζ.
If $F_1(v_0)$ depends on φ; we obtain a representation analogous to (*), only the g_{ks} have to be multiplied with the new resulting Fourier coefficients. A particular solution of the inhomogenous problem for w_k determined by the set-up method has always the property to decay exponentially for $\zeta \to \infty$ and to behave like $\frac{1}{k^{2m}}$ for $k \to \infty$ in correspondence with the regularity of the coefficients of F_1 and the $g_s(\varphi)$. Consequently, v_1 is also of classical boundary layer type.

3.3. A more general equation of second order

We are interested in a solution periodic in φ of

$$Lv \equiv -\frac{\partial^2 v}{\partial \zeta^2} + b\zeta \frac{\partial v}{\partial \zeta} + \frac{1}{\gamma(\varphi)} \frac{\partial v}{\partial \varphi} + cv = 0$$

$$v\big|_{\zeta=0} = g(\varphi)\,, \qquad 0 < \varphi < 2\pi\,,$$

where b and c are constants and c > 0.
Let again

$$v(\zeta,\varphi) = \sum_{k=-\infty}^{\infty} V_k(\zeta)\,\phi_k(\varphi)\,.$$

$V_k(\zeta)$ can immediately be expressed by parabolic cylinder functions (cf. A 1.3), it holds that

$$V_k(\zeta) = \sqrt{\tfrac{2}{\pi}}\, c_k \exp\left(\tfrac{b}{4}\zeta^2\right) \begin{cases} \dfrac{\Gamma\left(1-\frac{c+\mu_k}{2b}\right)}{2^{\frac{c+\mu_k}{b}-1}}\, D_{\frac{c+\mu_k}{b}-1}\left(\sqrt{-b}\,\zeta\right), & b<0 \\[2ex] \dfrac{\Gamma\left(\frac{1}{2}+\frac{c+\mu_k}{2b}\right)}{2^{-\frac{c+\mu_k}{b}}}\, D_{-\frac{c+\mu_k}{b}}\left(\sqrt{b}\,\zeta\right), & b>0\,, \end{cases}$$

where

$$c_k = \frac{1}{P}\int_0^{2\pi} \gamma(\varphi)\, g(\varphi)\, \phi_{-k}(\varphi)\, d\varphi\,.$$

The $V_k(\zeta)$ are uniformly bounded and for $\zeta \to \infty$ it holds

$$V_k(\zeta) \sim \begin{cases} d_1\, \zeta^{\frac{c+\mu_k}{b}-1} \exp\left(\frac{b}{2}\zeta^2\right), & b<0 \\ d_2\, \zeta^{-\frac{c+\mu_k}{b}} & b>0\,. \end{cases}$$

If $g(\varphi)$ is sufficiently regular, then the Fourier series for v and its derivatives converge uniformly.
A generalization on

$$Lv \equiv (-1)^m \frac{\partial^{2m} v}{\partial \zeta^{2m}} + b\zeta \frac{\partial v}{\partial \zeta} + \frac{1}{\gamma(\varphi)} \frac{\partial v}{\partial \varphi} + c\,v = 0$$

is connected with difficulties which arise because it is impossible to express explicitly the solution of the homogenous problems and from the difficult treatment of the inhomogenous problems (cf. A 1.3).

3.4. A special equation of second order

We are interested in a solution periodic in φ of

$$Lv \equiv -\frac{\partial^2 v}{\partial \zeta^2} + \frac{1}{\gamma(\varphi)} \frac{\partial v}{\partial \varphi} = 0$$

$$v\Big|_{\zeta=0} = g(\varphi) \ .$$

Setting again

$$v(\zeta,\varphi) = \sum_{k=-\infty}^{\infty} V_k(\zeta)\, \phi_k(\varphi)$$

with

$$\phi_k(\varphi) = \exp\Big(\mu_k \int_0^{\varphi} \gamma(\alpha)\, d\alpha\Big), \qquad \mu_k = \frac{2\pi k i}{p}, \qquad p = \int_0^{2\pi} \gamma(\alpha)\, d\alpha$$

the $V_k(\zeta)$ satisfy

$$-\frac{d^2 V_k}{d\zeta^2} + \mu_k V_k = 0$$

$$V_k\Big|_{\zeta=0} = g_k$$

with

$$g_k = \frac{1}{p} \int_0^{2\pi} \gamma(\varphi)\, g(\varphi)\, \phi_{-k}(\varphi)\, d\varphi \ .$$

From this it follows

$$V_0(\zeta) \equiv g_0$$

and

$$V_k(\zeta) = g_k \ \exp(\varkappa_k \zeta) \ , \qquad k \geq 1 \ ,$$

where $\varkappa_k$ denotes the root with negative real part of $\varkappa^2 = \mu_k$, consequently

$$\varkappa_k = \left(\frac{2\pi|k|}{P}\right)^{\frac{1}{2}} \begin{cases} -\frac{1}{2}\sqrt{2} - \frac{1}{2}\sqrt{2}\, i\ , & k > 0 \\ -\frac{1}{2}\sqrt{2} + \frac{1}{2}\sqrt{2}\, i\ , & k < 0\ . \end{cases}$$

If $g(\varphi)$ is sufficiently regular, then $v(\zeta,\varphi) - \dot{g}_0$ is a classical boundary layer function. This result (in the case of $\gamma(\varphi) \equiv 1$) can also be found in CHASMINSKIJ (1963).

A 4 Parabolic boundary layers (III)

4.1. A degenerating equation of second order (1)

We consider

$$L v \equiv -\frac{\partial^2 v}{\partial \zeta^2} + \zeta^\alpha \frac{\partial v}{\partial t} = 0$$

$$v\big|_{t=0} = 0$$

$$v\big|_{\zeta=0} = g(t)$$

where $t > 0$, $\zeta > 0$ and α is a positive constant.
Analogously, as in A 2.1, it is possible to express explicitly

$$v(\zeta,t) = \frac{1}{\Gamma(2+2\alpha)} \int\limits_{\frac{(2+2\alpha)^2 \zeta^{\frac{1}{2+2\alpha}}}{t}}^{\infty} \tau^{1+2\alpha} e^{-\tau} g\left(t - \frac{(2+2\alpha)^2 \zeta^{\frac{1}{2+2\alpha}}}{\tau}\right) d\tau\ .$$

v has also properties analogous to those of the solution of the boundary equation in A 2.1: If $g(t)$ is regular and $g(0) = 0$, then $v, \frac{\partial v}{\partial t}, \frac{\partial v}{\partial \zeta}, \frac{\partial^2 v}{\partial \zeta^2}$ are uniformly bounded classical boundary layer functions. However, $\frac{\partial^2 v}{\partial t^2}$ and $\frac{\partial^2 v}{\partial t \partial \zeta}$ are only then bounded if $g'(0) = 0$. For the present degenerating parabolic equation, in difference to A 2, it cannot be guaranteed a-priori that bounded solutions are uniquely determined.
Let

$$L\, w = 0$$

$$w\big|_{t=0} = \varphi(\zeta)$$

$$w\big|_{\zeta=0} = \psi(t)\ , \qquad \varphi(0) = \psi(0)\ .$$

Then it holds the following

<u>Lemma:</u> If $\varphi(\zeta)$ and $\psi(t)$ are continuous, then any bounded solution satisfies the estimate

$$\min(\inf\varphi, \inf\psi) \leq w \leq \max(\sup\varphi, \sup\psi) \text{ in } \bar{G} = [0,\infty) \times [0,T],$$

in particular, any bounded solution is uniquely determined. ∎

For the proof we first introduce the abbreviations

$$m = \min(\inf\varphi, \inf\psi), \qquad M = \max(\sup\varphi, \sup\psi).$$

Now let

$$w_1 = \mu \ln(1+\zeta) + M - w,$$

$$w_2 = \mu \ln(1+\zeta) + w - m, \qquad \mu > 0.$$

Since, according to the assumption, w is bounded, then for each μ there exists a $K(\mu)$ such that

$$w_{1,2} \geq 0 \qquad \text{in } [K(\mu), \infty) \times [0,T].$$

Furthermore, according to the construction of w_1, w_2, it holds

$$L w_{1,2} < 0, \qquad w_{1,2}\Big|_{\zeta=0} \geq 0, \qquad w_{1,2}\Big|_{t=0} \geq 0.$$

From the maximum principle for bounded domains (cf. Chapter 2) it then follows that

$$w_{1,2} \geq 0 \qquad \text{in } [0, K(\mu)] \times [0,T].$$

Summing up it holds

$$m - \mu \ln(1+\zeta) \leq w(\zeta,t) \leq M + \mu \ln(1+\zeta)$$

for all ζ, t. Now the asserted estimate follows from the free choice of μ.

4.2. A degenerating equation of second order (2)

We consider

$$Lv \equiv -\frac{\partial^2 v}{\partial \zeta^2} - \zeta^l \frac{\partial v}{\partial \zeta} + x\zeta^{l-1}\frac{\partial v}{\partial x} = 0$$

$$v\Big|_{\zeta=0} = g(x)$$

where $x \in (-1,+1)$ and $g(x) \in C^\infty(-1,+1)$.
Setting

$$v^*(\zeta,x) = x^k V^*(\zeta)$$

we are able to separate the variables and obtain

$$-\frac{d^2V^*}{d\zeta^2} - \zeta^l\frac{dV^*}{d\zeta} + k\zeta^{l-1}V^* = 0 .$$

Introducing

$$V^* = \exp\left(-\frac{\zeta^{l+1}}{l+1}\right)V \qquad and \qquad \frac{\zeta^{l+1}}{l+1} = z$$

then $V(z)$ satisfies the confluent hyper-geometric differential equation

$$z\frac{d^2V}{dz^2} + \left(\frac{l}{l+1} - z\right)\frac{dV}{dz} - \left(\frac{k}{l+1} + \frac{l}{l+1}\right)V = 0$$

which has a solution $\tilde{V}_k(z)$ with

$$\tilde{V}_k(z) = {}_1F_1\left(\frac{k}{l+1}+\frac{l}{l+1};\frac{l}{l+1};z\right) + \frac{\Gamma(-\frac{1}{l+1})\Gamma(\frac{k}{l+1}+1)}{\Gamma(\frac{1}{l+1})\Gamma(\frac{k}{l+1}+\frac{l}{l+1})} z^{\frac{1}{l+1}}\,{}_1F_1\left(\frac{k}{l+1}+1;1+\frac{1}{l+1};z\right).$$

It holds

$$\tilde{V}_k(0) = 1$$

and

$$\tilde{V}_k(z) \sim d\, z^{-\left(\frac{k}{l+1}+\frac{l}{l+1}\right)} \quad for\ \frac{\pi}{2} \geq \arg z > -\frac{3\pi}{2}\ and\ |z| \to \infty .$$

Consequently, $\tilde{V}_k\left(\frac{\zeta^{l+1}}{l+1}\right)\exp\left(-\frac{\zeta^{l+1}}{l+1}\right)$ is regular for any integer, non-negative k and a classical boundary layer function.
The solution of the above problems is

$$v(\zeta,x) = \sum_{k=0}^{\infty} g_k x^k \tilde{V}_k\left(\frac{\zeta^{l+1}}{l+1}\right)\exp\left(-\frac{\zeta^{l+1}}{l+1}\right).$$

BIBLIOGRAPHY

I. MONOGRAPHS AND PROCEEDINGS ON SINGULAR PERTURBATIONS

Application of the method of matched asymptotic expansions to boundary value problems for differential equations (1979). (Применение метода согласования асимптотических разложений к краевым задачам для дифференциальных уравнений), Свердловск, Изд. Уральского научного центра АН СССР.

BERG, L. (1968), Asymptotische Darstellungen und Entwicklungen. Berlin, Deutscher Verlag der Wissenschaften.

BLEISTEIN, N., HANDELSMAN, D. A. (1975), Asymptotic expansions of integrals. New York; Holt, Rinehat and Winston.

BRAUNER,C. M., GAY, B., MATHIEU, J. (eds.) (1977), Singular perturbations and boundary layer theory. Lect. Notes Math., 594.

DE GROEN, P. P. N. (1976), Singularly perturbed differential operators of second order. Amsterdam, Math. Centrum.

Difference methods for solving boundary value problems containing a small parameter and discontinuous boundary conditions (1976). (Разностные методы решения краевых задач с малым параметром и разрывными краевыми условиями), Свердловск, Изд. Уральского научного центра АН СССР.

Differential equations containing a small parameter (1980). (Дифференциальные уравнения с малым параметром), Свердловск, Изд. Уральского научного центра АН СССР.

DOOLAN, E. P., MILLER, J. J. H., SCHILDERS, W. H. A. (1980) Uniform numerical methods for problems with initial and boundary layers. Dublin, Boole press.

ECKHAUS, W. (1973), Matched asymptotic expansions and singular perturbations. Amsterdam, North Holland Publ. Comp.

ECKHAUS, W. (1979), Asymptotic analysis of singular perturbations, Amsterdam, North Holland Publ. Comp.

ERDELYI, A. (1956), Asymptotic expansions. Dover, New York.

GOERING, H. (1977), Asymptotische Methoden zur Lösung von Differentialgleichungen. Berlin, Akademie-Verlag

GRASMAN, J. (1971), On the birth of boundary layer, Amsterdam, Math. Centrum.

HEMKER, P. W., MILLER, J. J. H. (eds.), Numerical analysis of singular perturbation problem, New York, Academic press.

LIONS, J. L. (1973), Perturbation singulieres dans les problemes aux limites et en Contrôle Optimal. Lect. Notes Math., 323.

MILLER, J. J. H. (ed.) (1980), Boundary and interior layers - computional and asymptotic methods. Dublin, Boole press limited.

NAYFEH, A. H. (1973), Perturbation methods. New York, Wiley.

OLVER, F. W. J. (1974), Asymptotics and special functions. New York, Academic press.

O'MALLEY, R. E. (1974), Introduction to singular perturbations. New York, Academic press.

PITTNAUER, F. (1972), Vorlesungen über asymptotische Reihen. Lect. Notes Math. 301.

SIBUYA, Y. (1975), Global theory of second order linear ordinary differential equations with a polynomial coefficient, Amsterdam.

SIROVICH, L. (1971), Techniques of asymptotic analysis, New York, Springer-Verlag.

VASILEVA, A. V., BUTUZOV, V. F. (1973) (Васильева, А. В., Бутузов, В. Ф.), Асимптотические разложения решений сингулярно возмущённых уравнений, Москва, Наука.

VASILEVA, A. V., BUTUZOW, V. F. (1978) (Васильева, А. В., Бутузов, В. Ф.), Сингулярно возмущённые уравнения в критических случаях. Москва, Изд. Московского университета.

VAN DYKE, M. (1975), Perturbation methods in fluid mechanics, Stanford, Parabolic Press.

VENTCEL, A. D., FREIDLIN, M. I. (1979) (Вентцель, А. Д., Фрейдлин, М. И.), Флуктуации в динамических системах под действием малых случайных возмущений, Москва, Наука.

WASOW, W. (1965), Asymptotic expansions for ordinary differential equations. New York, Wiley.

COLE, J. D. (1968) Perturbation Methods in Applied Mathematics, Waltham, Mass., Blaisdell.

KEVORKIAN, J., COLE, J. D. (1981) Perturbation Methods in Applied Mathematics, New York, Berlin, Springer.

II. ARTICLES ON SINGULAR PERTURBATIONS

ACKERBERG, R. C., O'MALLEY, R. E. (1970), Boundary layer problems exhibiting resonance, Stud. Appl. Math., 49, 277-295.

ARONSON, D. G. (1956), Linear parabolic differential equations containing a small parameter, Journal Rat. Mech. a. Anal., 5, 1003-1014.

BARDOS, C., BRÉZIS, D., BREZIS, H. (1973), Perturbations singulieres et prolongements maximaux d' operateurs positives, Arch. Rat. Mech. Anal., 53, 1, 69-100.

BARTON, N. G. (1976), On the asymptotic solution of an elliptic interior layer problem, Journal Austral. Math. Soc., Ser. B 19, 493-512.

BERG, L. (1965), Asymptotische Lösungen für Operatorgleichungen. ZAMM 45, 333-352.

BERG, L. (1976), Two matching rules, Beitr. Anal. 9, 7-10.

BERGER, M. S., FRAENKEL, L. E. (1971), On singular perturbations of nonlinear operator equations. Indiana Univ. Math. J., 20, 7, 623-631.

BESJES, J. G. (1974), Singular perturbation problems for linear parabolic differential operators of arbitrary order. J. Math. Anal. Appl. 48, 594-609.

BESJES, J. G. (1975a), Singular perturbation problems for linear elliptic differential operators of arbitrary order I. Degeneration to elliptic operators. J. Math. Anal. Appl. 49, 24-46.

BESJES, J. G. (1975b), Singular perturbation problems for linear elliptic differential operators of arbitrary order II. Degeneration to first order operators. J. Math. Anal. Appl. 49, 324-346.

BOBISUD, L. E. (1968), Second order linear parabolic equations with a small parameter. Arch. Rat. Mech. Anal., 27, 385-397.

BUTUZOV, V. F. (1975), (Бутузов, В. Ф.), Об асимптотике решений сингулярно возмущённых уравнений эллиптического типа в прямоугольной области. Дифф. уравн., II, 6, I030-I04I.

BUTUZOV, V. F. (1977) (Бутузов, В. Ф.), О построении пограничных функций в некоторых сингулярно возмущённых задачах эллиптического типа. Дифф. уравн., I3, I0, I829-I835.

BUTUZOV, V. F. (1979) (Бутузов, В. Ф.), Угловой погранслой в сингулярно возмущённых задачах с частными производными. Дифф. уравн., I5, I0, I848-I862.

CHAZ'MINSKIJ, R. Z. (1963), (Хазьминский, Р. З.), Диффузионные процессы с малым параметром. Изв. АН СССР, сер. Мат., 27, 1281-1300.

COOK, P., ECKHAUS, W. (1973), Resonance in a boundary value problem of singular perturbation type. Stud. Appl. Math., 52, 2, 129-140.

DE GROEN, P. P. N. (1974/75), Elliptic singular perturbations of first order operators with critical points.
Proc. Royal Soc.Edinb., Sect. A 74, 91-113.

DE JAGER, E. M. (1972), Singular elliptic perturbations of vanishing first order differential operators. Lect. Notes Math. 280, 73-86.

DE RUIJTER, W. (1979), Boundary layers in large scale ocean circulation. Lect. Notes Math., 711, 125-145.

DEVINATZ, A.; ELLIS, R.; FRIEDMAN, A. (1974), The asymptotic behaviour of the first real eigenvalue of second order elliptic operators with a small parameter in the highest derivatives II.
Indiana Univ. Math. J., 23, 991-1011.

DEVINATZ, A., FRIEDMAN, A. (1978a), Asymptotic behaviour of the principal eigenfunction for a singularly perturbed Dirichlet problem.
Indiana Univ. Math. Journal, 27, 1, 143-157.

DEVINATZ, A., FRIEDMAN, A. (1978b), The asymptotic behaviour of the solution of a singularly perturbed Dirichlet problem.
Indiana Univ. Math. Journal, 27, 3, 527-537.

DORR, F. W. (1970), Some examples of singular perturbation problems with turning points. SIAM J. math. Anal. 1, 141-146.

ECKHAUS, W.; DE JAGER, E. M. (1966), Asymptotic solutions of singular perturbation problems for linear differential equations of elliptic type. Arch. Rat. Mech. Anal. 23, 1, 26-86.

ECKHAUS, W. (1969), On the foundations of the method of matched asymptotic approximations. J. de Mecanique 8, 2, 265-300.

ECKHAUS, W. (1977a), Formal approximations and singular perturbations. SIAM Review 19, No. 4, 593-633.

ECKHAUS, W. (1977b), Matching principles and composite expansions. Lect. Notes in Math. 594, Springer, Berlin-Heidelberg-New York, 146-177.

FELGENHAUER, A. (1980a), Topology of asymptotics; Proc. of the conf. topology and measure II, part. 1, Greifswald, 27-32.

FELGENHAUER, A. (1980b), Anwendung von H^1-Maximumprinzipien auf singulär gestörte Probleme zweiter Ordnung.
Wiss. Z. THM Otto von Guericke Magdeburg, 24, 5, 25-27.

FELGENHAUER, A. (1982), Verbindungsregeln (matching rules) für Projektoren in linearen Räumen, Zeitschr. f. Analysis u. Anwendungen 1.

FIFE, P. C. (1973), Semilinear elliptic boundary value problems with small parameters. Arch. Rat. Mech. Anal. 52, 3, 205-232.

FRAENKEL, L. E. (1969), On the method of matched asymptotic expansions I: A matching principle, II: Some applications of the composite series, III: Two boundary value problems; Proc. Cambridge Phil. Soc.65,209-284.

GÖHDE, D. (1981), Innere Abschätzungen bei singulär gestörten Randwertproblemen; Wissenschaftliche Beiträge der IHS Zwickau, 7, 2, 75-77.

GRASMAN, J. (1977), Elliptic singular perturbations of first order differential operators vanishing at an interior surface.
TW 169, Math. Centrum Amsterdam.

GRASMAN, J. (1979), Small random perturbations of dynamical systems with application in population genetics. Lect. Notes Math.,711, 158-175.

GRASMAN, J., MATKOWSY, B. J. (1977), A variational approach to singularly perturbed boundary value problems for ordinary and partial differential equations with turning points.
SIAM J. Appl. Math., 32, Nr. 3, 588.

HABETS, P. (1980), Singular perturbations of elliptic boundary value problems. Lect. Notes Math. 827, 115-130.

HINTZE, A. (1981), Ein singulär gestörtes elliptisches Eigenwertproblem für die dritte Randwertaufgabe. Dissertation, TH Magdeburg.

HOWES, F. A. (1979), Singularly perturbed semilinear elliptic boundary value problems. Comm. partial diff. equ. 4, 1, 1-39.

HOWES, F. A. (1979), Robin and Neumann problems for a class of singularly perturbed semilinear elliptic equations.
Journ. diff. equ., 34, 1, 55-73.

HOWES, F. A. (1980), Some two parameter problems for singularly perturbed equations of elliptic type. Indiana Univ. Math.J., 29,127-145.

ILJIN, A. M.; GORKOV, Ju. P.; LELIKOVA, E. F. (1975) (Ильин, А. М., Горьков, Ю. П., Леликова, Е. Ф.), Асимптотика решения эллиптического уравнения с малым параметром при старших производных в окрестности особой характеристики предельного уравнения. Труды семинара им. И. Г. Петровского, 75-133.

ILJIN, A. M., LELIKOVA, E. F. (1975) (Ильин, А. М., Леликова, Е. Ф.), О методе сращивания для уравнения $\varepsilon \Delta u - a(x,y)u_y = f$. Мат. сборник, 96(138), 568-583.

ISAKOVA, E. K. (1966), (Исакова, Е. К.), Асимптотическое разложение решения параболического уравнения с малым параметром. Мат. сборник, 69(111), 300-320.

KAMIN, S. (1978), Elliptic perturbations of a first order operator with a singular point of attracting type. Indiana Univ. Math. J., 27, 6, 935-952.

KAMIN, S. (1979), On elliptic singular perturbation problems with turning points. SIAM J. Math. Anal., 10, 3, 447-455.

KAMIN, S. (1979), On elliptic equations with a small parameter in the highest derivatives. Comm. part. diff. equ., 4, 573-593.

KAPLUN, SAUL (1967), Fluid mechanics and singular perturbations (ed. by P. A. Lagerstrom, L. N. Howard, Ching Shilin). Acad. Press, NY-London.

KELLEY, W. G. (1981), The Dirichlet problem for singularly perturbed quasilinear elliptic equations. J. diff. equ. 40, 1, 37-52.

KIFER, Y. (1980), On the principal eigenvalue in a singular perturbation problem with hyperbolic limit points and circles. Journal diff. equ., 37, 1, 108-133.

LAGERSTROM, Casten (1972), Baic concepts underlying singular perturbationstechniques. SIAM Review 14, 1, 63-120.

LELIKOVA, E. F. (1976), (Леликова, Е. Ф.), Об асимптотическом решении эллиптического уравнения второго порядка с малым параметром при старшей производной. Дифф. уравн., 12, 10, 1852-1865.

LELIKOVA, E. F. (1978) (Леликова, Е. Ф.), Метод сращивания асимптотических разложений для уравнения $\varepsilon \Delta u - a u_z = f$ в параллелепипеде. Дифф. уравн., 14, 9, 1638-1648.

LELIKOVA, E. F. (1979) (Леликова, Е. Ф.), Об асимптотике решения эллиптического уравнения с малым параметром в области с кусочно гладкой границей, in: Application of the method of matched asymptotic...,40-57.

LUBE, G. (1980), Die asymptotische Lösung einer Klasse singulär gestörter Randwertprobleme zweiter Ordnung. Dissertation, TH Magdeburg.

LUBE, G.; TOBISKA, L. (1980), Singulär gestörte elliptische Probleme beliebiger Ordnung mit Typänderung. Math. Nachr. 98, 123-134.

MANGEL, M. (1979), Small fluctuations in systems with multiple steady states. SIAM J. appl. math., 36, 3, 544-572.

MANGEL, M. (1980), Small fluctuations in systems with multiple limit cycles. SIAM J. appl. math., 38, 1, 120-138.

MATKOWSKY, B. J., SCHUSS, Z. (1976), On the problem of exit. Bull. Americ. Math. Soc., 82, 321-324.

MATKOWSKY, B. J., SCHUSS, Z. (1977), The exit problem for randomly perturbed dynamical systems. SIAM J. appl. math., 33, 2, 365-382.

MAUSS, J. (1973), On first order matching process for singular functions, in: Spectral theory and asymptotics of diff. eq. (ed.: de Jager), Amsterdam, London, 163-173.

MAUSS, J. (1979), On matching principles. Lect. Notes in Math., 711, Berlin-Heidelberg-New York, 1-8.

OLVER, F. W. J. (1978), Sufficient conditions for Ackerberg-O'Malley resonance. SIAM J. math. Anal., 9, 2, 328-355.

REINHARD, H.-J. (1978), On nonlinear singular and regular perturbation problems. Preprint, Frankfurt/M.

ROOS, H.-G. (1978), Ein singulär gestörtes Problem für eine lineare parabolische Differentialgleichung beliebiger Ordnung. Comm. math. univ. carl. 19, 13-25.

ROOS, H.-G. (1979), Die Konstruktion asymptotischer Fundamentalsysteme für lineare Differentialgleichungen mit Wendepunkten. Comm. math. univ. carolinae, 20, 195-205.

SHAIGARDANOV, JU. Z. (1979), (Шайгарданов, Ю. З.), Об асомптотическом решении первой краевой задачи для эллиптического уравнения четвертого порядка с малым параметром при старшей производной, in: Application of the method of matched asymptotic expansions..., 58-72.

SCHUSS, Z., MATKOWSKY, B. J. (1979), The exit problem: a new approach to diffusion across potential barriers. SIAM J. appl. math., 36, 3, 604-623.

SHIVAMOGGI, B. K. (1978), Method of matched asymptotic expansions - asymptotic matching principle for higher approximations. ZAMM 58, 354-356.

STENGLE, G. (1971), A new basis for uniform asymptotic solutions of differential equations containing one or several parameters. Trans. AMS 156, 1-43.

SWANSON, C. A., SCHULZER, M. (1961), Asymptotic solutions of equations in Banach space. Canad. J. Math. 13, 493-504.

TOBISKA, L. (1981), A-priori Abschätzungen für singulär gestörte elliptische Probleme 2. Ordnung. Beiträge zur Analysis 17, 41-48.

VAN HARTEN, A. (1976), On an elliptic singular perturbation problem. Lect. Notes Math., 564, 485-495.

VAN HARTEN, A. (1978), Nonlinear singular perturbation problems: Proofs of correctness of a formal approximation based on a contraction principle in a Banach space. J. Math. Anal. Appl. 65, 126-168.

VENTZEL, A. D. (1975), (Вентцель, А. Д.), Об асимптотике первого собственного значения дифференциального оператора второго порядка с малым параметром при старшей производной, Теор. вероят. и прим.,20,3,6I0-6I3.

WILLIAMS, R. G. (1981), The problem of stochastic exit. SIAM J. appl. math., 40, 2, 208-233.

FRIEDMAN, A. (1973), The asymptotic behaviour of the first real eigenvalue of a second order elliptic operator with a small parameter in the highest derivatives. Indiana Univ. Math. J., 22, 1005-1015.

III. ADDITIONAL REFERENCES

AGMON, S., DOUGLIS, A. and NIRENBERG, N. (1959), Estimates near the boundary for solutions of elliptic partial differential equations satisfying general boundary conditions I.
Comm. Pure Appl. Math. 12, 623-727.

ARNOLD, V. I. (1979), Gewöhnliche Differentialgleichungen.
Moskau 1975, Nauka. German Translation: Berlin 1979, VEB Deutscher Verlag der Wissenschaften.

AZZAM, A. (1980), On Differentiability properties of solutions of elliptic differential equations. J. Math. Anal. Appl. 75, 431-440.

BARDOS, C. (1970), Problemes aux limites pour les equations aux derivés partielles du les ordre, théoreme d'approximation et application a l'equation de transport. Ann. Sc. E.N.S., 3, 185-233.

BATEMAN, G., ERDELYI, A. (1974), (Бейтмен, Г., Эрдейи, А.), Высшие трансцендентные Функции II. Москва (Transl. from engl.).

EIDEL'MAN, S. D. (1956), (Эйделман, С. Д.), О фундаментальных решениях параболических систем. Мат. сборник, 38(80), I, 5I-92.

FRIEDMAN, A. (1964), Partial differential equations of parabolic type Englewood Cliffs.

FRIEDRICHS, K. O. (1958), Symmetric positive linear differential equations. Comm. Pure Appl. Math., 11, 333-418.

KAMKE, E. (1965), Differentialgleichungen, Lösungsmethoden und Lösungen II. Leipzig, Geest und Portig K.-G.

KÖTHE, G. (1960), Topologische lineare Räume.
Berlin, Göttingen, Heidelberg; Springer.

LEFSCHETZ, S. (1957), Differential equations: Geometric theory.
New York, Intersc. Publ., Inc.

MICHAEL, J. H. (1977), A general theory for linear elliptic partial differential equations. J. diff. equat. 23, 1-29.

MICHAILOV, V. P. (1960), (Михайлов, В. П.), Решение смешанной задачи для параболических систем методом потенциалов. ДАН СССР, I32,2,29I-294.

MIRANDA, C. (1970), Partial differential equations of elliptic type.
Berlin.

OLEJNIK, O. A., RADKEVIC, E. U. (1973), Second order equations with nonnegative characteristic form. English transl.: Americ. Math. soc., New York.

PROTTER, M. H., WEINBERGER, H. F. (1967), Maximum principles in differential equations. Englewood Cliffs, New Jersey, (Prentice-Hall,Inc.)

ROMASKEVIC, L. P. (1981), (Ромаскевич, Л. П.), О классических решениях квазилинейных вырождающихся эллиптических уравнений второго порядка с разрывными коэффициентами. Успехи мат. наук, 36, 2(208), 205-206.

SOLONNIKOV, V. A. (1965), (Солонников, В. А.), О краевых задачах для линейных параболических систем дифференциальных уравнений общего вида. Труды мат. института им. Стеклова, 83.

STROACK, D. W., VARADHAN, S. R. S. (1971), Diffusion processes with boundary conditions. Comm. Pure Appl. Math., 24, 147-225.

YAMADA, H. (1979), Boundary value problems for real linear partial differential equations of first order. Kodai Math. J., 2, 346-361.

SUBJECT INDEX

Bei Fragen zur Produktsicherheit wenden Sie sich bitte an:
If you have any questions regarding product safety,
please contact:

Walter de Gruyter GmbH
Genthiner Straße 13
10785 Berlin
productsafety@degruyterbrill.com